Principles of Condensed Matter Physics

Principles of Condensed Matter Physics

Edited by
Jessica Finley

Larsen & Keller
www.larsen-keller.com

Principles of Condensed Matter Physics
Edited by Jessica Finley
ISBN: 978-1-63549-707-6 (Hardback)

Larsen & Keller
Published by Larsen and Keller Education,
5 Penn Plaza,
19th Floor,
New York, NY 10001, USA

Cataloging-in-Publication Data

Principles of condensed matter physics / edited by Jessica Finley.
p. cm.
Includes bibliographical references and index.
ISBN 978-1-63549-707-6
1. Condensed matter. 2. Physics. I. Finley, Jessica.
QC173.454 .P75 2018
530.41--dc23

The publisher's policy is to use permanent paper from mills that operate a sustainable forestry policy. Furthermore, the publisher ensures that the text paper and cover boards used have met acceptable environmental accreditation standards.

Printed and bound in China.

For more information regarding Larsen and Keller Education and its products, please visit the publisher's website www.larsen-keller.com

Table of Contents

Preface

The branch of physics which uses three basic laws namely electromagnetism, statistical mechanics and quantum mechanics, to understand the physical and behavioral properties of condensed matter is known as condensed matter physics. Condensed matter physics uses the elements of nanotechnology, chemistry, biophysics, materials science, etc. It has application in the areas of laser technology, quantum computation and semiconductor devices among many others. This book attempts to understand the multiple branches that fall under this discipline. Most of the topics introduced in it cover new techniques and the applications of the subject. This textbook attempts to assist those with a goal of delving into this complex field.

A foreword of all Chapters of the book is provided below:

Chapter 1 - Condensed matter theory is a discipline of physics that explores and studies the physical properties of condensed matter. The theory includes the laws of electromagnetism, statistical mechanics, and quantum mechanics. The section also briefly discusses other concepts of statistical physics. This chapter on statistical physics offers an insightful focus, keeping in mind the complex subject matter; **Chapter 2 -** Phase is a themodynamic system within which all the physical states of matter are uniform. Superfluidity is a characteristic wherin a liquid has zero viscosity and hence, loses no kinetic energy when flowing. The chapter strategically encompasses and incorporates the major components and key concepts of condensed matter physics, providing a complete understanding; **Chapter 3 -** Phase transition describes the change in state of matter into solid, liquid, gas and rarely, plasma. The physical properties of matter become uniform during phase transitions. Condensed matter theory is best understood in confluence with the major topics listed in the following chapter; **Chapter 4 -** Various theories related to condensed matter physics include Bloch wave, Ginzburg–Landau theory, Fermi gas, etc. Fermi gas contains a large number of fermions and is a phase of matter. The topics discussed in the chapter are of great importance to broaden the existing knowledge on condensed matter physics; **Chapter 5 -** Crystallography studies the arrangement of atoms in crystalline solids. X-rays, electron and neutron beams are used to study crystalline structure. The aspects elucidated in this chapter are of vital importance, and provide a better understanding of condensed matter theory.

I would like to thank the entire editorial team who made sincere efforts for this book and my family who supported me in my efforts of working on this book. I take this opportunity to thank all those who have been a guiding force throughout my life.

Editor

1

Understanding Condensed Matter Physics

Condensed matter theory is a discipline of physics that explores and studies the physical properties of condensed matter. The theory includes the laws of electromagnetism, statistical mechanics, and quantum mechanics. The section also briefly discusses other concepts of statistical physics. This chapter on statistical physics offers an insightful focus, keeping in mind the complex subject matter.

Condensed Matter Physics

Condensed matter physics is a branch of physics that deals with the physical properties of condensed phases of matter, where particles adhere to each other. Condensed matter physicists seek to understand the behavior of these phases by using physical laws. In particular, they include the laws of quantum mechanics, electromagnetism and statistical mechanics.

The most familiar condensed phases are solids and liquids while more exotic condensed phases include the superconducting phase exhibited by certain materials at low temperature, the ferromagnetic and antiferromagnetic phases of spins on crystal lattices of atoms, and the Bose–Einstein condensate found in ultracold atomic systems. The study of condensed matter physics involves measuring various material properties via experimental probes along with using methods of theoretical physics to develop mathematical models that help in understanding physical behavior.

The diversity of systems and phenomena available for study makes condensed matter physics the most active field of contemporary physics: one third of all American physicists self-identify as condensed matter physicists, and the Division of Condensed Matter Physics is the largest division at the American Physical Society. The field overlaps with chemistry, materials science, and nanotechnology, and relates closely to atomic physics and biophysics. The theoretical physics of condensed matter shares important concepts and methods with that of particle physics and nuclear physics.

A variety of topics in physics such as crystallography, metallurgy, elasticity, magnetism, etc., were treated as distinct areas until the 1940s, when they were grouped together as *solid state physics*. Around the 1960s, the study of physical properties of liquids was added to this list, forming the basis for the new, related specialty of condensed matter physics. According to physicist Philip Warren Anderson, the term was coined by him and Volker Heine, when they changed the name of their group at the Cavendish Laboratories, Cambridge from *Solid state theory* to *Theory of Condensed Matter* in 1967, as they felt it did not exclude their interests in the study of liquids, nuclear matter, and so on. Although Anderson and Heine helped popularize the name "condensed matter", it had been present in Europe for some years, most prominently in the form of a journal published in English, French, and German by Springer-Verlag titled *Physics of Condensed Matter*, which was launched in 1963. The funding environment and Cold War politics of the 1960s and 1970s were also factors that lead some physicists to prefer the name "condensed matter physics", which emphasized the commonality of scientific problems encountered by physicists working on solids, liquids, plasmas, and other complex matter, over "solid state physics", which was often associated with the industrial applications of metals and semiconductors. The Bell Telephone Laboratories was one of the first institutes to conduct a research program in condensed matter physics.

References to "condensed" state can be traced to earlier sources. For example, in the introduction to his 1947 book *Kinetic Theory of Liquids*, Yakov Frenkel proposed that "The kinetic theory of liquids must accordingly be developed as a generalization and extension of the kinetic theory of solid bodies. As a matter of fact, it would be more correct to unify them under the title of 'condensed bodies'".

History

Classical Physics

Heike Kamerlingh Onnes and Johannes van der Waals with the helium *liquefactor* in Leiden (1908)

One of the first studies of condensed states of matter was by English chemist Humphry Davy, in the first decades of the nineteenth century. Davy observed that of the forty chemical elements known at the time, twenty-six had metallic properties such as lustre, ductility and high electrical and thermal conductivity. This indicated that the atoms in John Dalton's atomic theory were not indivisible as Dalton claimed, but had inner structure. Davy further claimed that elements that were then believed to be gases, such as nitrogen and hydrogen could be liquefied under the right conditions and would then behave as metals.

In 1823, Michael Faraday, then an assistant in Davy's lab, successfully liquefied chlorine and went on to liquefy all known gaseous elements, except for nitrogen, hydrogen, and oxygen. Shortly after, in 1869, Irish chemist Thomas Andrews studied the phase transition from a liquid to a gas and coined the term critical point to describe the condition where a gas and a liquid were indistinguishable as phases, and Dutch physicist Johannes van der Waals supplied the theoretical framework which allowed the prediction of critical behavior based on measurements at much higher temperatures. By 1908, James Dewar and Heike Kamerlingh Onnes were successfully able to liquefy hydrogen and then newly discovered helium, respectively.

Paul Drude in 1900 proposed the first theoretical model for a classical electron moving through a metallic solid. Drude's model described properties of metals in terms of a gas of free electrons, and was the first microscopic model to explain empirical observations such as the Wiedemann–Franz law. However, despite the success of Drude's free electron model, it had one notable problem: it was unable to correctly explain the electronic contribution to the specific heat and magnetic properties of metals, and the temperature dependence of resistivity at low temperatures.

In 1911, three years after helium was first liquefied, Onnes working at University of Leiden discovered superconductivity in mercury, when he observed the electrical resistivity of mercury to vanish at temperatures below a certain value. The phenomenon completely surprised the best theoretical physicists of the time, and it remained unexplained for several decades. Albert Einstein, in 1922, said regarding contemporary theories of superconductivity that "with our far-reaching ignorance of the quantum mechanics of composite systems we are very far from being able to compose a theory out of these vague ideas".

Advent of Quantum Mechanics

Drude's classical model was augmented by Wolfgang Pauli, Arnold Sommerfeld, Felix Bloch and other physicists. Pauli realized that the free electrons in metal must obey the Fermi–Dirac statistics. Using this idea, he developed the theory of paramagnetism in 1926. Shortly after, Sommerfeld incorporated the Fermi–Dirac statistics into the free electron model and made it better able to explain the heat capacity. Two years later, Bloch used quantum mechanics to describe the motion of a quantum electron in a periodic lattice. The mathematics of crystal structures developed by Auguste Bravais, Yevgraf Fyodorov and others was used to classify crystals by their symmetry group, and tables of crystal structures were the basis for the series *International Tables of Crystallography*, first published in 1935. Band structure calculations was first used in 1930 to predict the properties of new materials, and in 1947 John Bardeen, Walter Brattain and William Shockley developed the first semiconductor-based transistor, heralding a revolution in electronics.

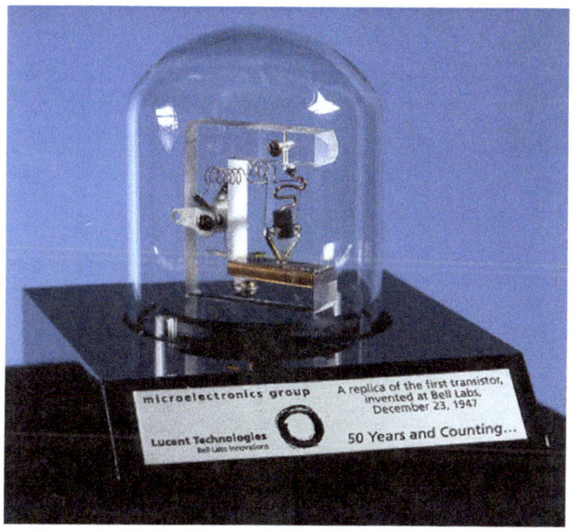

A replica of the first point-contact transistor in Bell labs

In 1879, Edwin Herbert Hall working at the Johns Hopkins University discovered a voltage developing across conductors transverse to an electric current in the conductor and magnetic field perpendicular to the current. This phenomenon arising due to the nature of charge carriers in the conductor came to be termed the Hall effect, but it was not properly explained at the time, since the electron was experimentally discovered 18 years later. After the advent of quantum mechanics, Lev Landau in 1930 developed the theory of landau quantization and laid the foundation for the theoretical explanation for the quantum Hall effect discovered half a century later.

Magnetism as a property of matter has been known in China since 4000 BC. However, the first modern studies of magnetism only started with the development of electrodynamics by Faraday, Maxwell and others in the nineteenth century, which included classifying materials as ferromagnetic, paramagnetic and diamagnetic based on their response to magnetization. Pierre Curie studied the dependence of magnetization on temperature and discovered the Curie point phase transition in ferromagnetic materials. In 1906, Pierre Weiss introduced the concept of magnetic domains to explain the main properties of ferromagnets. The first attempt at a microscopic description of magnetism was by Wilhelm Lenz and Ernst Ising through the Ising model that described magnetic materials as consisting of a periodic lattice of spins that collectively acquired magnetization. The Ising model was solved exactly to show that spontaneous magnetization cannot occur in one dimension but is possible in higher-dimensional lattices. Further research such as by Bloch on spin waves and Néel on antiferromagnetism led to developing new magnetic materials with applications to magnetic storage devices.

Modern Many-body Physics

A magnet levitating above a high-temperature superconductor. Today some physicists are working to understand high-temperature superconductivity using the AdS/CFT correspondence.

The Sommerfeld model and spin models for ferromagnetism illustrated the successful application of quantum mechanics to condensed matter problems in the 1930s. However, there still were several unsolved problems, most notably the description of superconductivity and the Kondo effect. After World War II, several ideas from quantum field theory were applied to condensed matter problems. These included recognition of collective excitation modes of solids and the important notion of a quasiparticle. Russian physicist Lev Landau used the idea for the Fermi liquid theory wherein low energy properties of interacting fermion systems were given in terms of what are now termed Landau-quasiparticles. Landau also developed a mean field theory for continuous phase transitions, which described ordered phases as spontaneous breakdown of symmetry. The theory also introduced the notion of an order parameter to distinguish between ordered phases. Eventually in 1965, John Bardeen, Leon Cooper and John Schrieffer developed the so-called BCS theory of superconductivity, based on the discovery that arbitrarily small attraction between two electrons of opposite spin mediated by phonons in the lattice can give rise to a bound state called a Cooper pair.

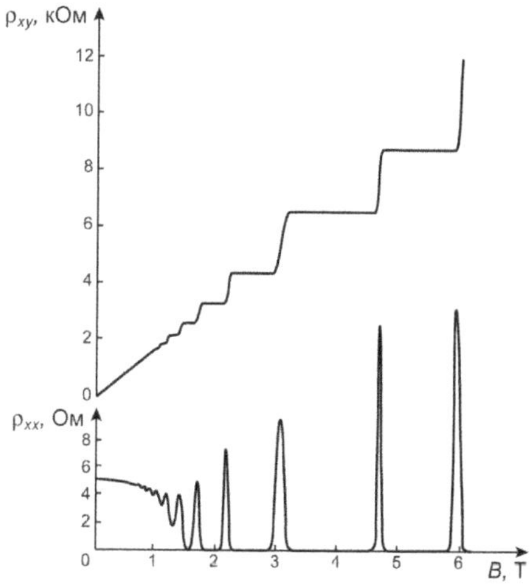

The quantum Hall effect: Components of the Hall resistivity as a function of the external magnetic field

The study of phase transition and the critical behavior of observables, termed critical phenomena, was a major field of interest in the 1960s. Leo Kadanoff, Benjamin Widom and Michael Fisher developed the ideas of critical exponents and widom scaling. These ideas were unified by Kenneth G. Wilson in 1972, under the formalism of the renormalization group in the context of quantum field theory.

The quantum Hall effect was discovered by Klaus von Klitzing in 1980 when he observed the Hall conductance to be integer multiples of a fundamental constant e^2 / h.. The effect was observed to be independent of parameters such as system size and impurities. In 1981, theorist Robert Laughlin proposed a theory explaining the unanticipated precision of the integral plateau. It also implied that the Hall conductance can be characterized in terms of a topological invariable called Chern number. Shortly after, in 1982, Horst Störmer and Daniel Tsui observed the fractional quantum Hall effect where the conductance was now a rational multiple of a constant. Laughlin, in 1983, realized that this was a consequence of quasiparticle interaction in the Hall states and formulated a variational method solution, named the Laughlin wavefunction. The study of topological properties of the fractional Hall effect remains an active field of research.

In 1986, Karl Müller and Johannes Bednorz discovered the first high temperature superconductor, a material which was superconducting at temperatures as high as 50 Kelvin. It was realized that the high temperature superconductors are examples of strongly correlated materials where the electron–electron interactions play an important role. A satisfactory theoretical description of high-temperature superconductors is still not known and the field of strongly correlated materials continues to be an active research topic.

In 2009, David Field and researchers at Aarhus University discovered spontaneous electric fields when creating prosaic films of various gases. This has more recently expanded to form the research area of spontelectrics.

In 2012 several groups released preprints which suggest that samarium hexaboride has the properties of a topological insulator in accord with the earlier theoretical predictions. Since samarium hexaboride is an established Kondo insulator, i.e. a strongly correlated electron material, the existence of a topological surface state in this material would lead to a topological insulator with strong electronic correlations.

Theoretical

Theoretical condensed matter physics involves the use of theoretical models to understand properties of states of matter. These include models to study the electronic properties of solids, such as the Drude model, the Band structure and the density functional theory. Theoretical models have also been developed to study the physics of phase transitions, such as the Ginzburg–Landau theory, critical exponents and the use of mathematical methods of quantum field theory and the renormalization group. Modern theoretical studies involve the use of numerical computation of electronic structure and mathematical tools to understand phenomena such as high-temperature superconductivity, topological phases, and gauge symmetries.

Emergence

Theoretical understanding of condensed matter physics is closely related to the notion of emergence, wherein complex assemblies of particles behave in ways dramatically different from their individual constituents. For example, a range of phenomena related to high temperature superconductivity are understood poorly, although the microscopic physics of individual electrons and lattices is well known. Similarly, models of condensed matter systems have been studied where collective excitations behave like photons and electrons, thereby describing electromagnetism as an emergent phenomenon. Emergent properties can also occur at the interface between materials: one example is the lanthanum aluminate-strontium titanate interface, where two non-magnetic insulators are joined to create conductivity, superconductivity, and ferromagnetism.

Electronic Theory of Solids

The metallic state has historically been an important building block for studying properties of solids. The first theoretical description of metals was given by Paul Drude in 1900 with the Drude model, which explained electrical and thermal properties by describing a metal as an ideal gas of then-newly discovered electrons. He was able to derive the empirical Wiedemann-Franz law and get results in close agreement with the experiments. This classical model was then improved by Arnold Sommerfeld who incorporated the Fermi–Dirac statistics of electrons and was able to explain the anomalous behavior of the specific heat of metals in the Wiedemann–Franz law. In 1912, The structure of crystalline solids was studied by Max von Laue and Paul Knipping, when they observed the X-ray diffraction pattern of crystals, and concluded that crystals get their structure from periodic lattices of atoms. In 1928, Swiss physicist Felix Bloch provided a wave function solution to the Schrödinger equation with a periodic potential, called the Bloch wave.

Calculating electronic properties of metals by solving the many-body wavefunction is often computationally hard, and hence, approximation methods are needed to obtain meaningful predictions. The Thomas–Fermi theory, developed in the 1920s, was used to estimate system energy and electronic density by treating the local electron density as a variational parameter. Later in the 1930s, Douglas Hartree, Vladimir Fock and John Slater developed the so-called Hartree–Fock wavefunction as an improvement over the Thomas–Fermi model. The Hartree–Fock method accounted for exchange statistics of single particle electron wavefunctions. In general, it's very difficult to solve the Hartree–Fock equation. Only the free electron gas case can be solved exactly. Finally in 1964–65, Walter Kohn, Pierre Hohenberg and Lu Jeu Sham proposed the density functional theory which gave realistic descriptions for bulk and surface properties of metals. The density functional theory (DFT) has been widely used since the 1970s for band structure calculations of variety of solids.

Symmetry Breaking

Some states of matter exhibit symmetry breaking, where the relevant laws of physics possess some symmetry that is broken. A common example is crystalline solids, which break continuous translational symmetry. Other examples include magnetized ferromagnets, which break rotational symmetry, and more exotic states such as the ground state of a BCS superconductor, that breaks U(1) phase rotational symmetry.

Goldstone's theorem in quantum field theory states that in a system with broken continuous symmetry, there may exist excitations with arbitrarily low energy, called the Goldstone bosons. For example, in crystalline solids, these correspond to phonons, which are quantized versions of lattice vibrations.

Phase Transition

Phase transition refers to the change of phase of a system, which is brought about by change in an external parameter such as temperature. Classical phase transition occurs at finite temperature when the order of the system was destroyed. For example, when ice melts and becomes water, the ordered crystal structure is destroyed. In quantum phase transitions, the temperature is set to absolute zero, and the non-thermal control parameter, such as pressure or magnetic field, causes the phase transitions when order is destroyed by quantum fluctuations originating from the Heisenberg uncertainty principle. Here, the different quantum phases of the system refer to distinct ground states of the Hamiltonian. Understanding the behavior of quantum phase transition is important in the difficult tasks of explaining the properties of rare-earth magnetic insulators, high-temperature superconductors, and other substances.

Two classes of phase transitions occur: first-order transitions and continuous transitions. For the later, the two phases involved do not co-exist at the transition temperature, also called critical point. Near the critical point, systems undergo critical behavior, wherein several of their properties such as correlation length, specific heat, and magnetic susceptibility diverge exponentially. These critical phenomena poses serious challenges to physicists because normal macroscopic laws are no longer valid in the region and novel ideas and methods must be invented to find the new laws that can describe the system.

The simplest theory that can describe continuous phase transitions is the Ginzburg–Landau theory, which works in the so-called mean field approximation. However, it can only roughly explain continuous phase transition for ferroelectrics and type I superconductors which involves long range microscopic interactions. For other types of systems that involves short range interactions near the critical point, a better theory is needed.

Near the critical point, the fluctuations happen over broad range of size scales while the feature of the whole system is scale invariant. Renormalization group methods successively average out the shortest wavelength fluctuations in stages while retaining their effects into the next stage. Thus, the changes of a physical system as viewed at different size scales can be investigated systematically. The methods, together with powerful computer simulation, contribute greatly to the explanation of the critical phenomena associated with continuous phase transition.

Experimental

Experimental condensed matter physics involves the use of experimental probes to try to discover new properties of materials. Such probes include effects of electric and magnetic fields, measuring response functions, transport properties and thermometry. Commonly used experimental methods include spectroscopy, with probes such as X-rays, infrared light and inelastic neutron scattering; study of thermal response, such as specific heat and measuring transport via thermal and heat conduction.

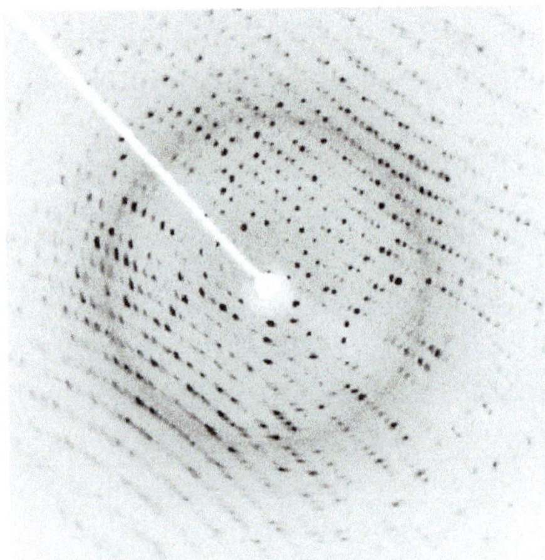

Image of X-ray diffraction pattern from a protein crystal

Scattering

Several condensed matter experiments involve scattering of an experimental probe, such as X-ray, optical photons, neutrons, etc., on constituents of a material. The choice of scattering probe depends on the observation energy scale of interest. Visible light has energy on the scale of 1 electron volt (eV) and is used as a scattering probe to measure variations in material properties such as dielectric constant and refractive index. X-rays have energies of the order of 10 keV and hence are able to probe atomic length scales, and are used to measure variations in electron charge density.

Neutrons can also probe atomic length scales and are used to study scattering off nuclei and electron spins and magnetization (as neutrons have spin but no charge). Coulomb and Mott scattering measurements can be made by using electron beams as scattering probes. Similarly, positron annihilation can be used as an indirect measurement of local electron density. Laser spectroscopy is an excellent tool for studying the microscopic properties of a medium, for example, to study forbidden transitions in media with nonlinear optical spectroscopy.

External Magnetic Fields

In experimental condensed matter physics, external magnetic fields act as thermodynamic variables that control the state, phase transitions and properties of material systems. Nuclear magnetic resonance (NMR) is a method by which external magnetic fields are used to find resonance modes of individual electrons, thus giving information about the atomic, molecular, and bond structure of their neighborhood. NMR experiments can be made in magnetic fields with strengths up to 60 Tesla. Higher magnetic fields can improve the quality of NMR measurement data. Quantum oscillations is another experimental method where high magnetic fields are used to study material properties such as the geometry of the Fermi surface. High magnetic fields will be useful in experimentally testing of the various theoretical predictions such as the quantized magnetoelectric effect, image magnetic monopole, and the half-integer quantum Hall effect.

Cold Atomic Gases

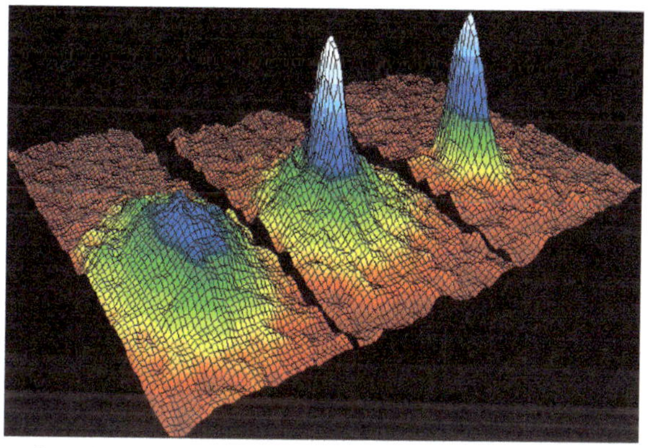

The first Bose–Einstein condensate observed in a gas of ultracold rubidium atoms. The blue and white areas represent higher density.

Ultracold atom trapping in optical lattices is an experimental tool commonly used in condensed matter physics, and in atomic, molecular, and optical physics. The method involves using optical lasers to form an interference pattern, which acts as a *lattice*, in which ions or atoms can be placed at very low temperatures. Cold atoms in optical lattices are used as *quantum simulators*, that is, they act as controllable systems that can model behavior of more complicated systems, such as frustrated magnets. In particular, they are used to engineer one-, two- and three-dimensional lattices for a Hubbard model with pre-specified parameters, and to study phase transitions for antiferromagnetic and spin liquid ordering.

In 1995, a gas of rubidium atoms cooled down to a temperature of 170 nK was used to experimentally realize the Bose–Einstein condensate, a novel state of matter originally predicted by S. N. Bose and Albert Einstein, wherein a large number of atoms occupy one quantum state.

Applications

Research in condensed matter physics has given rise to several device applications, such as the development of the semiconductor transistor, laser technology, and several phenomena studied in the context of nanotechnology. Methods such as scanning-tunneling microscopy can be used to control processes at the nanometer scale, and have given rise to the study of nanofabrication.

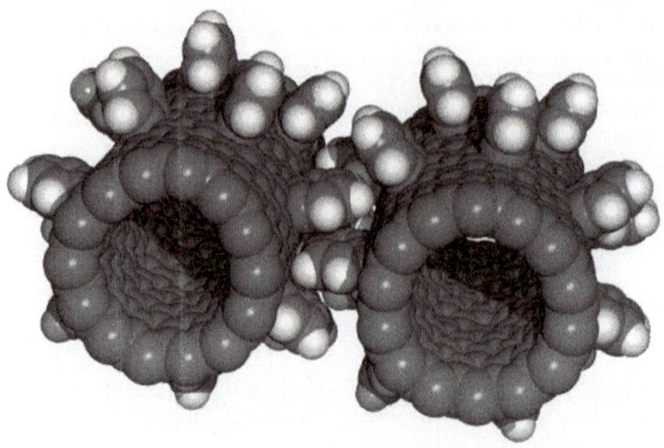

Computer simulation of *nanogears* made of fullerene molecules.
It is hoped that advances in nanoscience will lead to machines working on the molecular scale.

In quantum computation, information is represented by quantum bits, or qubits. The qubits may decohere quickly before useful computation is completed. This serious problem must be solved before quantum computing may be realized. To solve this problem, several promising approaches are proposed in condensed matter physics, including superconducting Josephson junction qubits, spintronic qubits using the spin orientation of magnetic materials, or the topological non-Abelian anyons from fractional quantum Hall effect states.

Condensed matter physics also has important uses for biophysics, for example, the experimental method of magnetic resonance imaging, which is widely used in medical diagnosis.

State of Matter

In physics, a state of matter is one of the distinct forms that matter takes on. Four states of matter are observable in everyday life: solid, liquid, gas, and plasma. Many other states are known to exist only in extreme situations, such as Bose–Einstein condensates, neutron-degenerate matter, and quark-gluon plasma, which only occur in situations of extreme cold, extreme density, and extremely high-energy color-charged matter respectively. Some other states are believed to be possible but remain theoretical for now.

Historically, the distinction is made based on qualitative differences in properties. Matter in the solid state maintains a fixed volume and shape, with component particles (atoms, molecules or ions) close together and fixed into place. Matter in the liquid state maintains a fixed volume, but has a variable shape that adapts to fit its container. Its particles are still close together but move freely. Matter in the gaseous state has both variable volume and shape, adapting both to fit its container. Its particles are neither close together nor fixed in place. Matter in the plasma state has variable volume and shape, but as well as neutral atoms, it contains a significant number of ions and electrons, both of which can move around freely. Plasma is the most common form of visible matter in the universe.

The four fundamental states of matter. Clockwise from top left, they are solid, liquid, plasma, and gas, represented by an ice sculpture, a drop of water, electrical arcing from a tesla coil, and the air around clouds, respectively.

The term phase is sometimes used as a synonym for state of matter, but a system can contain several immiscible phases of the same state of matter.

The Four Fundamental States

Solid

A crystalline solid: atomic resolution image of strontium titanate. Brighter atoms are Sr and darker ones are Ti

In a solid, the particles (ions, atoms, or molecules) are closely packed together. The forces between particles are strong so that the particles cannot move freely but can only vibrate. As a result, a solid has a stable, definite shape, and a definite volume. Solids can only change their shape by force, as when broken or cut.

In crystalline solids, the particles (atoms, molecules, or ions) are packed in a regularly ordered, repeating pattern. There are various different crystal structures, and the same substance can have more than one structure (or solid phase). For example, iron has a body-centred cubic structure at temperatures below 912 °C, and a face-centred cubic structure between 912 and 1394 °C. Ice has fifteen known crystal structures, or fifteen solid phases, which exist at various temperatures and pressures.

Glasses and other non-crystalline, amorphous solids without long-range order are not thermal equilibrium ground states; therefore they are described below as nonclassical states of matter.

Solids can be transformed into liquids by melting, and liquids can be transformed into solids by freezing. Solids can also change directly into gases through the process of sublimation, and gases can likewise change directly into solids through deposition.

Liquid

Structure of a classical monatomic liquid. Atoms have many nearest neighbors in contact, yet no long-range order is present.

A liquid is a nearly incompressible fluid that conforms to the shape of its container but retains a (nearly) constant volume independent of pressure. The volume is definite if the temperature and pressure are constant. When a solid is heated above its melting point, it becomes liquid, given that the pressure is higher than the triple point of the substance. Intermolecular (or interatomic or interionic) forces are still important, but the molecules have enough energy to move relative to each other and the structure is mobile. This means that the shape of a liquid is not definite but is determined by its container. The volume is usually greater than that of the corresponding solid, the best known exception being water, H_2O. The highest temperature at which a given liquid can exist is its critical temperature.

Gas

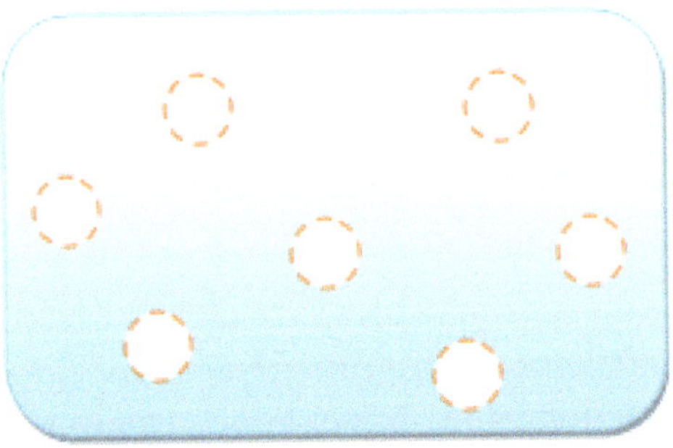

The spaces between gas molecules are very big. Gas molecules have very weak or no bonds at all. The molecules in "gas" can move freely and fast.

A gas is a compressible fluid. Not only will a gas conform to the shape of its container but it will also expand to fill the container.

In a gas, the molecules have enough kinetic energy so that the effect of intermolecular forces is small (or zero for an ideal gas), and the typical distance between neighboring molecules is much greater than the molecular size. A gas has no definite shape or volume, but occupies the entire container in which it is confined. A liquid may be converted to a gas by heating at constant pressure to the boiling point, or else by reducing the pressure at constant temperature.

At temperatures below its critical temperature, a gas is also called a vapor, and can be liquefied by compression alone without cooling. A vapor can exist in equilibrium with a liquid (or solid), in which case the gas pressure equals the vapor pressure of the liquid (or solid).

A supercritical fluid (SCF) is a gas whose temperature and pressure are above the critical temperature and critical pressure respectively. In this state, the distinction between liquid and gas disappears. A supercritical fluid has the physical properties of a gas, but its high density confers solvent properties in some cases, which leads to useful applications. For example, supercritical carbon dioxide is used to extract caffeine in the manufacture of decaffeinated coffee.

Plasma

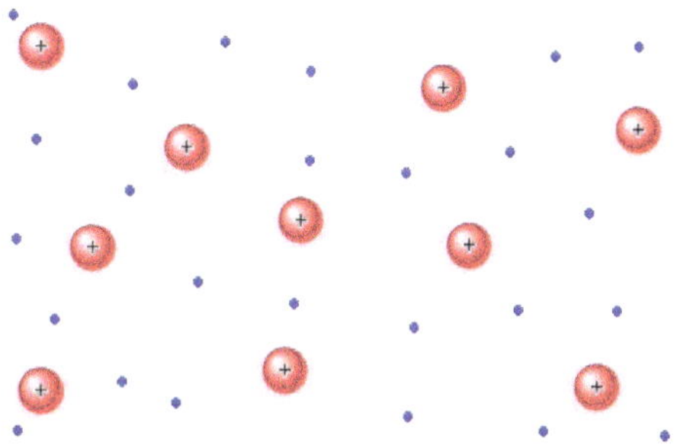

In a plasma, electrons are ripped away from their nuclei, forming an electron "sea". This gives it the ability to conduct electricity.

Like a gas, plasma does not have definite shape or volume. Unlike gases, plasmas are electrically conductive, produce magnetic fields and electric currents, and respond strongly to electromagnetic forces. Positively charged nuclei swim in a "sea" of freely-moving disassociated electrons, similar to the way such charges exist in conductive metal. In fact it is this electron "sea" that allows matter in the plasma state to conduct electricity.

The plasma state is often misunderstood, but it is actually quite common on Earth, and the majority of people observe it on a regular basis without even realizing it. Lightning, electric sparks, fluorescent lights, neon lights, plasma televisions, some types of flame and the stars are all examples of illuminated matter in the plasma state.

A gas is usually converted to a plasma in one of two ways, either from a huge voltage difference between two points, or by exposing it to extremely high temperatures.

Heating matter to high temperatures causes electrons to leave the atoms, resulting in the presence of free electrons. At very high temperatures, such as those present in stars, it is assumed that essentially all electrons are "free", and that a very high-energy plasma is essentially bare nuclei swimming in a sea of electrons.

Phase Transitions

A state of matter is also characterized by phase transitions. A phase transition indicates a change in structure and can be recognized by an abrupt change in properties. A distinct state of matter can be defined as any set of states distinguished from any other set of states by a phase transition. Water can be said to have several distinct solid states. The appearance of superconductivity is associated with a phase transition, so there are superconductive states. Likewise, ferromagnetic states are demarcated by phase transitions and have distinctive properties. When the change of state occurs in stages the intermediate steps are called mesophases. Such phases have been exploited by the introduction of liquid crystal technology.

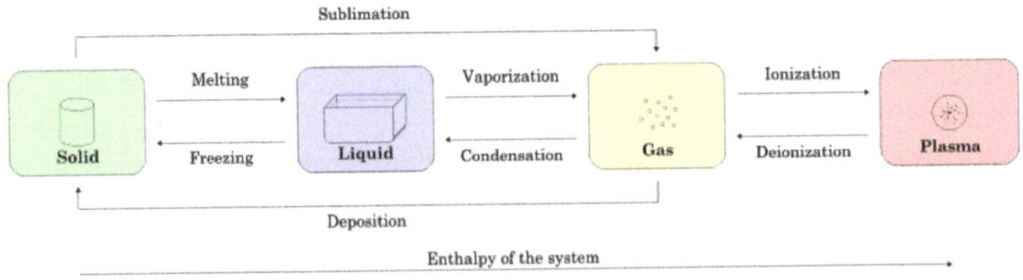

This diagram illustrates transitions between the four fundamental states of matter

The state or *phase* of a given set of matter can change depending on pressure and temperature conditions, transitioning to other phases as these conditions change to favor their existence; for example, solid transitions to liquid with an increase in temperature. Near absolute zero, a substance exists as a solid. As heat is added to this substance it melts into a liquid at its melting point, boils into a gas at its boiling point, and if heated high enough would enter a plasma state in which the electrons are so energized that they leave their parent atoms.

Forms of matter that are not composed of molecules and are organized by different forces can also be considered different states of matter. Superfluids (like Fermionic condensate) and the quark–gluon plasma are examples.

In a chemical equation, the state of matter of the chemicals may be shown as (s) for solid, (l) for liquid, and (g) for gas. An aqueous solution is denoted (aq). Matter in the plasma state is seldom used (if at all) in chemical equations, so there is no standard symbol to denote it. In the rare equations that plasma is used in plasma is symbolized as (p).

Non-classical States

Glass

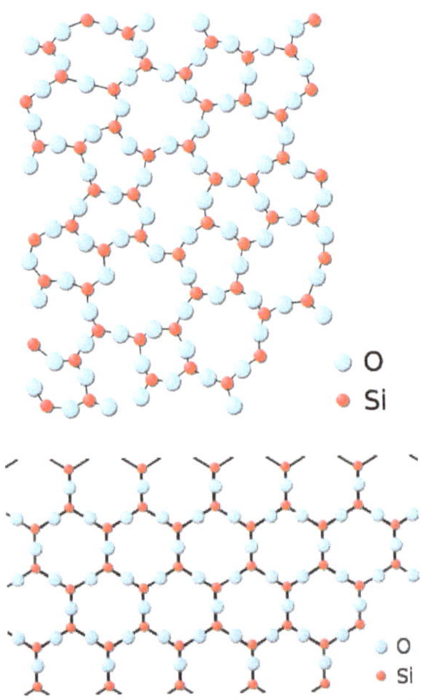

Schematic representation of a random-network glassy form (left) and ordered crystalline lattice (right) of identical chemical composition.

Glass is a non-crystalline or amorphous solid material that exhibits a glass transition when heated towards the liquid state. Glasses can be made of quite different classes of materials: inorganic networks (such as window glass, made of silicate plus additives), metallic alloys, ionic melts, aqueous solutions, molecular liquids, and polymers. Thermodynamically, a glass is in a metastable state with respect to its crystalline counterpart. The conversion rate, however, is practically zero.

Crystals with Some Degree of Disorder

A plastic crystal is a molecular solid with long-range positional order but with constituent molecules retaining rotational freedom; in an orientational glass this degree of freedom is frozen in a quenched disordered state.

Similarly, in a spin glass magnetic disorder is frozen.

Liquid Crystal States

Liquid crystal states have properties intermediate between mobile liquids and ordered solids. Generally, they are able to flow like a liquid, but exhibiting long-range order. For example, the nematic phase consists of long rod-like molecules such as para-azoxyanisole, which is nematic in the temperature range 118–136 °C. In this state the molecules flow as in a liquid, but they all point in the same direction (within each domain) and cannot rotate freely. Like a crystalline solid, but unlike a liquid, liquid crystals react to polarized light.

Several types have technological importance, for example, in liquid crystal displays.

Magnetically Ordered

Transition metal atoms often have magnetic moments due to the net spin of electrons that remain unpaired and do not form chemical bonds. In some solids the magnetic moments on different atoms are ordered and can form a ferromagnet, an antiferromagnet or a ferrimagnet.

In a ferromagnet—for instance, solid iron—the magnetic moment on each atom is aligned in the same direction (within a magnetic domain). If the domains are also aligned, the solid is a permanent magnet, which is magnetic even in the absence of an external magnetic field. The magnetization disappears when the magnet is heated to the Curie point, which for iron is 768 °C.

An antiferromagnet has two networks of equal and opposite magnetic moments, which cancel each other out so that the net magnetization is zero. For example, in nickel(II) oxide (NiO), half the nickel atoms have moments aligned in one direction and half in the opposite direction.

In a ferrimagnet, the two networks of magnetic moments are opposite but unequal, so that cancellation is incomplete and there is a non-zero net magnetization. An example is magnetite (Fe_3O_4), which contains Fe^{2+} and Fe^{3+} ions with different magnetic moments.

A quantum spin liquid (QSL) is a disordered state in a system of interacting quantum spins which preserves its disorder to very low temperatures, unlike other disordered states. It is not a liquid in physical sense, but a solid whose magnetic order is inherently disordered. The name "liquid" is due to an analogy with the molecular disorder in a conventional liquid. A QSL is neither a ferromagnet, where magnetic domains are parallel, nor an antiferromagnet, where the magnetic domains are antiparallel; instead, the magnetic domains are randomly oriented. This can be realized e.g. by geometrically frustrated magnetic moments that cannot point uniformly parallel or antiparallel. When cooling down and settling to a state, the domain must "choose" an orientation, but if the possible states are similar in energy, one will be chosen randomly. Consequently, despite strong short-range order, there is no long-range magnetic order.

Microphase-separated

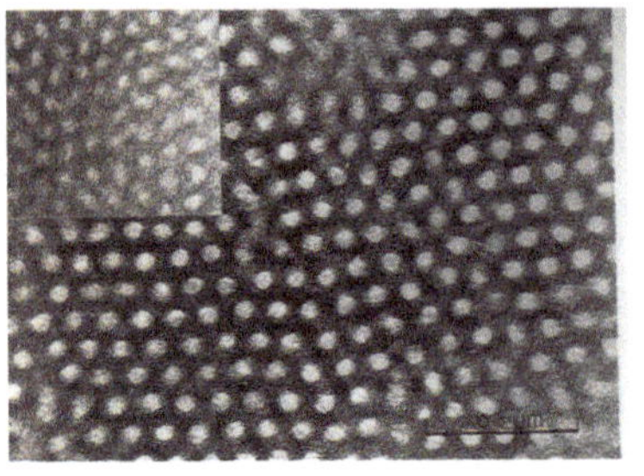

SBS block copolymer in TEM

Copolymers can undergo microphase separation to form a diverse array of periodic nanostructures, as shown in the example of the styrene-butadiene-styrene block copolymer shown. Microphase separation can be understood by analogy to the phase separation between oil and water. Due to chemical incompatibility between the blocks, block copolymers undergo a similar phase separation. However, because the blocks are covalently bonded to each other, they cannot demix macroscopically as water and oil can, and so instead the blocks form nanometer-sized structures. Depending on the relative lengths of each block and the overall block topology of the polymer, many morphologies can be obtained, each its own phase of matter.

Ionic liquids also display microphase separation. The anion and cation are not necessarily compatible and would demix otherwise, but electric charge attraction prevents them from separating. Their anions and cations appear to diffuse within compartmentalized layers or micelles instead of freely as in a uniform liquid.

Low-temperature States

Superfluid

Liquid helium in a superfluid phase creeps up on the walls of the cup in a Rollin film, eventually dripping out from the cup.

Close to absolute zero, some liquids form a second liquid state described as superfluid because it has zero viscosity (or infinite fluidity; i.e., flowing without friction). This was discovered in 1937 for helium, which forms a superfluid below the lambda temperature of 2.17 K. In this state it will attempt to "climb" out of its container. It also has infinite thermal conductivity so that no temperature gradient can form in a superfluid. Placing a superfluid in a spinning container will result in quantized vortices.

These properties are explained by the theory that the common isotope helium-4 forms a Bose–Einstein condensate in the superfluid state. More recently, Fermionic condensate superfluids have been formed at even lower temperatures by the rare isotope helium-3 and by lithium-6.

Bose–Einstein Condensate

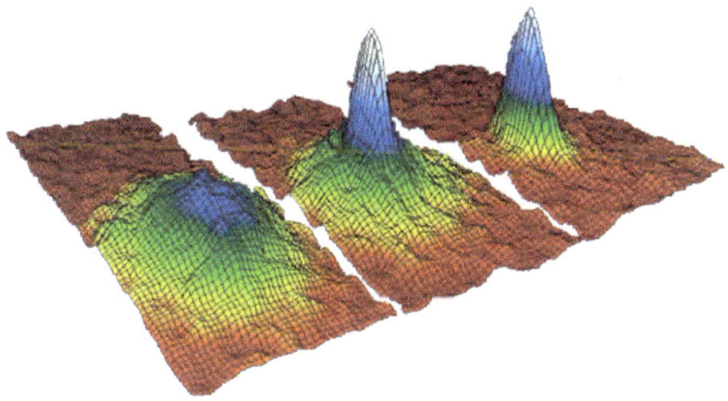

Velocity in a gas of rubidium as it is cooled: the starting material is on the left, and Bose–Einstein condensate is on the right.

In 1924, Albert Einstein and Satyendra Nath Bose predicted the "Bose–Einstein condensate" (BEC), sometimes referred to as the fifth state of matter. In a BEC, matter stops behaving as independent particles, and collapses into a single quantum state that can be described with a single, uniform wavefunction.

In the gas phase, the Bose–Einstein condensate remained an unverified theoretical prediction for many years. In 1995, the research groups of Eric Cornell and Carl Wieman, of JILA at the University of Colorado at Boulder, produced the first such condensate experimentally. A Bose–Einstein condensate is "colder" than a solid. It may occur when atoms have very similar (or the same) quantum levels, at temperatures very close to absolute zero (−273.15 °C).

Fermionic Condensate

A *fermionic condensate* is similar to the Bose–Einstein condensate but composed of fermions. The Pauli exclusion principle prevents fermions from entering the same quantum state, but a pair of fermions can behave as a boson, and multiple such pairs can then enter the same quantum state without restriction.

Rydberg Molecule

One of the metastable states of strongly non-ideal plasma is Rydberg matter, which forms upon condensation of excited atoms. These atoms can also turn into ions and electrons if they reach a certain temperature. In April 2009, *Nature* reported the creation of Rydberg molecules from a Rydberg atom and a ground state atom, confirming that such a state of matter could exist. The experiment was performed using ultracold rubidium atoms.

Quantum Hall State

A *quantum Hall state* gives rise to quantized Hall voltage measured in the direction perpendicular to the current flow. A *quantum spin Hall state* is a theoretical phase that may pave the way for the development of electronic devices that dissipate less energy and generate less heat. This is a derivation of the Quantum Hall state of matter.

Photonic Matter

Photonic matter is a phenomenon where photons interacting with a gas develop apparent mass, and can interact with each other, even forming photonic "molecules". The source of mass is the gas, which is massive. This is in contrast to photons moving in empty space, which have no rest mass, and cannot interact.

Dropleton

A "quantum fog" of electrons and holes that flow around each other and even ripple like a liquid, rather than existing as discrete pairs.

High-energy States

Degenerate Matter

Under extremely high pressure, as in the cores of dead stars, ordinary matter undergoes a transition to a series of exotic states of matter collectively known as degenerate matter, which are supported mainly by quantum mechanical effects. In physics, "degenerate" refers to two states that have the same energy and are thus interchangeable. Degenerate matter is supported by the Pauli exclusion principle, which prevents two fermionic particles from occupying the same quantum state. Unlike regular plasma, degenerate plasma expands little when heated, because there are simply no momentum states left. Consequently, degenerate stars collapse into very high densities. More massive degenerate stars are smaller, because the gravitational force increases, but pressure does not increase proportionally.

Electron-degenerate matter is found inside white dwarf stars. Electrons remain bound to atoms but are able to transfer to adjacent atoms. Neutron-degenerate matter is found in neutron stars. Vast gravitational pressure compresses atoms so strongly that the electrons are forced to combine with protons via inverse beta-decay, resulting in a superdense conglomeration of neutrons. Normally free neutrons outside an atomic nucleus will decay with a half life of just under 15 minutes, but in a neutron star, the decay is overtaken by inverse decay. Cold degenerate matter is also present in planets such as Jupiter and in the even more massive brown dwarfs, which are expected to have a core with metallic hydrogen. Because of the degeneracy, more massive brown dwarfs are not significantly larger. In metals, the electrons can be modeled as a degenerate gas moving in a lattice of non-degenerate positive ions.

Quark Matter

In regular cold matter, quarks, fundamental particles of nuclear matter, are confined by the strong force into hadrons that consist of 2–4 quarks, such as protons and neutrons. Quark matter or quantum chromodynanamical (QCD) matter is a group of phases where the strong force is overcome and quarks are deconfined and free to move. Quark matter phases occur at extremely high densities or temperatures, and there are no known ways to produce them in equilibrium in the laboratory; in ordinary conditions, any quark matter formed immediately undergoes radioactively decay.

Strange matter is a type of quark matter that is suspected to exist inside some neutron stars close to the Tolman–Oppenheimer–Volkoff limit (approximately 2–3 solar masses), although there is no direct evidence of its existence. In strange matter, part of the energy available manifests as strange quarks, a heavier analogue of the common down quark. It may be stable at lower energy states once formed, although this is not known.

Quark–gluon plasma is a very high-temperature phase in which quarks become free and able to move independently, rather than being perpetually bound into particles, in a sea of gluons, subatomic particles that transmit the strong force that binds quarks together. This is analogous to the liberation of electrons from atoms in a plasma. This state is briefly attainable in extremely high-energy heavy ion collisions in particle accelerators, and allows scientists to observe the properties of individual quarks, and not just theorize. Quark–gluon plasma was discovered at CERN in 2000. Unlike plasma, which flows like a gas, interactions within QGP are strong and it flows like a liquid.

At high densities but relatively low temperatures, quarks are theorized to form a quark liquid whose nature is presently unknown. It forms a distinct color-flavor locked (CFL) phase at even higher densities. This phase is superconductive for color charge. These phases may occur in neutron stars but they are presently theoretical.

Color-glass Condensate

Color-glass condensate is a type of matter theorized to exist in atomic nuclei traveling near the speed of light. According to Einstein's theory of relativity, a high-energy nucleus appears length contracted, or compressed, along its direction of motion. As a result, the gluons inside the nucleus appear to a stationary observer as a "gluonic wall" traveling near the speed of light. At very high energies, the density of the gluons in this wall is seen to increase greatly. Unlike the quark–gluon plasma produced in the collision of such walls, the color-glass condensate describes the walls themselves, and is an intrinsic property of the particles that can only be observed under high-energy conditions such as those at RHIC and possibly at the Large Hadron Collider as well.

Very High Energy States

The gravitational singularity predicted by general relativity to exist at the center of a black hole is *not* a phase of matter; it is not a material object at all (although the mass-energy of matter contributed to its creation) but rather a property of spacetime at a location. It could be argued, of course, that all particles are properties of spacetime at a location, leaving a half-note of controversy on the subject.

Other Proposed States

Supersolid

A supersolid is a spatially ordered material (that is, a solid or crystal) with superfluid properties. Similar to a superfluid, a supersolid is able to move without friction but retains a rigid shape. Although a supersolid is a solid, it exhibits so many characteristic properties different from other solids that many argue it is another state of matter.

String-net Liquid

In a string-net liquid, atoms have apparently unstable arrangement, like a liquid, but are still consistent in overall pattern, like a solid. When in a normal solid state, the atoms of matter align themselves in a grid pattern, so that the spin of any electron is the opposite of the spin of all electrons touching it. But in a string-net liquid, atoms are arranged in some pattern that requires some electrons to have neighbors with the same spin. This gives rise to curious properties, as well as supporting some unusual proposals about the fundamental conditions of the universe itself.

Superglass

A superglass is a phase of matter characterized, at the same time, by superfluidity and a frozen amorphous structure.

Dark Matter

While dark matter is estimated to comprise 83% of the mass of matter in the universe, most of its properties remain a mystery due to the fact that it neither absorbs nor emits electromagnetic radiation, and there are many competing theories regarding what dark matter is actually made of. Thus, while it is hypothesized to exist and comprise the vast majority of matter in the universe, almost all of its properties are unknown and a matter of speculation, because it has only been observed through its gravitational effects.

Solid

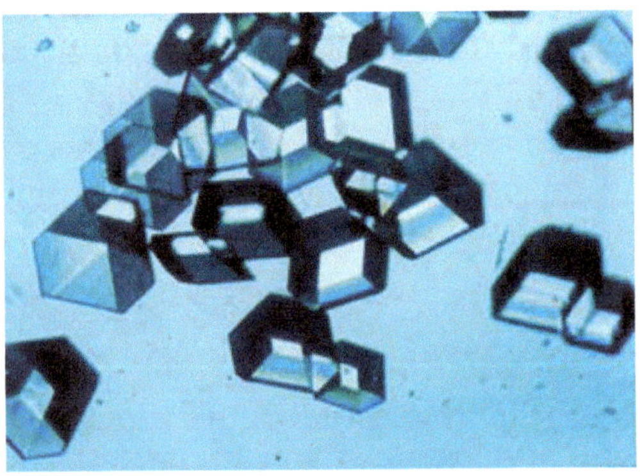

Single crystalline form of solid insulin

Solid is one of the four fundamental states of matter (the others being liquid, gas, and plasma). It is characterized by structural rigidity and resistance to changes of shape or volume. Unlike a liquid, a solid object does not flow to take on the shape of its container, nor does it expand to fill the entire volume available to it like a gas does. The atoms in a solid are tightly bound to each other, either in a regular geometric lattice (crystalline solids, which include metals and ordinary ice) or irregularly (an amorphous solid such as common window glass).

The branch of physics that deals with solids is called solid-state physics, and is the main branch of condensed matter physics (which also includes liquids). Materials science is primarily concerned with the physical and chemical properties of solids. Solid-state chemistry is especially concerned with the synthesis of novel materials, as well as the science of identification and chemical composition.

Microscopic Description

Model of closely packed atoms within a crystalline solid

The atoms, molecules or ions which make up solids may be arranged in an orderly repeating pattern, or irregularly. Materials whose constituents are arranged in a regular pattern are known as crystals. In some cases, the regular ordering can continue unbroken over a large scale, for example diamonds, where each diamond is a single crystal. Solid objects that are large enough to see and handle are rarely composed of a single crystal, but instead are made of a large number of single crystals, known as crystallites, whose size can vary from a few nanometers to several meters. Such materials are called polycrystalline. Almost all common metals, and many ceramics, are polycrystalline.

In other materials, there is no long-range order in the position of the atoms. These solids are known as amorphous solids; examples include polystyrene and glass.

Whether a solid is crystalline or amorphous depends on the material involved, and the conditions in which it was formed. Solids which are formed by slow cooling will tend to be crystalline, while solids which are frozen rapidly are more likely to be amorphous. Likewise, the specific crystal structure adopted by a crystalline solid depends on the material involved and on how it was formed.

While many common objects, such as an ice cube or a coin, are chemically identical throughout, many other common materials comprise a number of different substances packed together. For example, a typical rock is an aggregate of several different minerals and mineraloids, with no specific chemical composition. Wood is a natural organic material consisting primarily of cellulose fibers embedded in a matrix of organic lignin. In materials science, composites of more than one constituent material can be designed to have desired properties.

Classes of Solids

The forces between the atoms in a solid can take a variety of forms. For example, a crystal of sodium chloride (common salt) is made up of ionic sodium and chlorine, which are held together by ionic bonds. In diamond or silicon, the atoms share electrons and form covalent bonds. In metals, electrons are shared in metallic bonding. Some solids, particularly most organic compounds, are held together with van der Waals forces resulting from the polarization of the electronic charge cloud on each molecule. The dissimilarities between the types of solid result from the differences between their bonding.

Metals

The pinnacle of New York's Chrysler Building, the world's tallest steel-supported brick building, is clad with stainless steel.

Metals typically are strong, dense, and good conductors of both electricity and heat. The bulk of the elements in the periodic table, those to the left of a diagonal line drawn from boron to polonium, are metals. Mixtures of two or more elements in which the major component is a metal are known as alloys.

People have been using metals for a variety of purposes since prehistoric times. The strength and reliability of metals has led to their widespread use in construction of buildings and other structures, as well as in most vehicles, many appliances and tools, pipes, road signs and railroad tracks. Iron and aluminium are the two most commonly used structural metals, and they are also the most abundant metals in the Earth's crust. Iron is most commonly used in the form of an alloy, steel, which contains up to 2.1% carbon, making it much harder than pure iron.

Because metals are good conductors of electricity, they are valuable in electrical appliances and for carrying an electric current over long distances with little energy loss or dissipation. Thus, electrical power grids rely on metal cables to distribute electricity. Home electrical systems, for example, are wired with copper for its good conducting properties and easy machinability. The high thermal conductivity of most metals also makes them useful for stovetop cooking utensils.

The study of metallic elements and their alloys makes up a significant portion of the fields of solid-state chemistry, physics, materials science and engineering.

Metallic solids are held together by a high density of shared, delocalized electrons, known as "metallic bonding". In a metal, atoms readily lose their outermost ("valence") electrons, forming positive ions. The free electrons are spread over the entire solid, which is held together firmly by electrostatic interactions between the ions and the electron cloud. The large number of free electrons gives metals their high values of electrical and thermal conductivity. The free electrons also prevent transmission of visible light, making metals opaque, shiny and lustrous.

More advanced models of metal properties consider the effect of the positive ions cores on the delocalised electrons. As most metals have crystalline structure, those ions are usually arranged into a periodic lattice. Mathematically, the potential of the ion cores can be treated by various models, the simplest being the nearly free electron model.

Minerals

A collection of various minerals

Minerals are naturally occurring solids formed through various geological processes under high pressures. To be classified as a true mineral, a substance must have a crystal structure with uniform physical properties throughout. Minerals range in composition from pure elements and simple salts to very complex silicates with thousands of known forms. In contrast, a rock sample is a random aggregate of minerals and/or mineraloids, and has no specific chemical composition. The vast majority of the rocks of the Earth's crust consist of quartz (crystalline SiO_2), feldspar, mica, chlorite, kaolin, calcite, epidote, olivine, augite, hornblende, magnetite, hematite, limonite and a few other minerals. Some minerals, like quartz, mica or feldspar are common, while others have been found in only a few locations worldwide. The largest group of minerals by far is the silicates (most rocks are ≥95% silicates), which are composed largely of silicon and oxygen, with the addition of ions of aluminium, magnesium, iron, calcium and other metals.

Ceramics

Ceramic solids are composed of inorganic compounds, usually oxides of chemical elements. They are chemically inert, and often are capable of withstanding chemical erosion that occurs in an acidic or caustic environment. Ceramics generally can withstand high temperatures ranging from 1000 to 1600 °C (1800 to 3000 °F). Exceptions include non-oxide inorganic materials, such as nitrides, borides and carbides.

Si_3N_4 ceramic bearing parts

Traditional ceramic raw materials include clay minerals such as kaolinite, more recent materials include aluminium oxide (alumina). The modern ceramic materials, which are classified as advanced ceramics, include silicon carbide and tungsten carbide. Both are valued for their abrasion resistance, and hence find use in such applications as the wear plates of crushing equipment in mining operations.

Most ceramic materials, such as alumina and its compounds, are formed from fine powders, yielding a fine grained polycrystalline microstructure which is filled with light scattering centers comparable to the wavelength of visible light. Thus, they are generally opaque materials, as opposed to transparent materials. Recent nanoscale (e.g. sol-gel) technology has, however, made possible the production of polycrystalline transparent ceramics such as transparent alumina and alumina compounds for such applications as high-power lasers. Advanced ceramics are also used in the medicine, electrical and electronics industries.

Ceramic engineering is the science and technology of creating solid-state ceramic materials, parts and devices. This is done either by the action of heat, or, at lower temperatures, using precipitation reactions from chemical solutions. The term includes the purification of raw materials, the study and production of the chemical compounds concerned, their formation into components, and the study of their structure, composition and properties.

Mechanically speaking, ceramic materials are brittle, hard, strong in compression and weak in shearing and tension. Brittle materials may exhibit significant tensile strength by supporting a static load. Toughness indicates how much energy a material can absorb before mechanical failure, while fracture toughness (denoted K_{Ic}) describes the ability of a material with inherent microstructural flaws to resist fracture via crack growth and propagation. If a material has a large value of fracture toughness, the basic principles of fracture mechanics suggest that it will most likely undergo ductile fracture. Brittle fracture is very characteristic of most ceramic and glass-ceramic materials which typically exhibit low (and inconsistent) values of K_{Ic}.

For an example of applications of ceramics, the extreme hardness of zirconia is utilized in the manufacture of knife blades, as well as other industrial cutting tools. Ceramics such as alumina, boron carbide and silicon carbide have been used in bulletproof vests to repel large-caliber rifle fire. Silicon nitride parts are used in ceramic ball bearings, where their high hardness makes them wear resistant. In general, ceramics are also chemically resistant and can be used in wet environments where steel bearings would be susceptible to oxidation (or rust).

As another example of ceramic applications, in the early 1980s, Toyota researched production of an adiabatic ceramic engine with an operating temperature of over 6000 °F (3300 °C). Ceramic engines do not require a cooling system and hence allow a major weight reduction and therefore greater fuel efficiency. In a conventional metallic engine, much of the energy released from the fuel must be dissipated as waste heat in order to prevent a meltdown of the metallic parts. Work is also being done in developing ceramic parts for gas turbine engines. Turbine engines made with ceramics could operate more efficiently, giving aircraft greater range and payload for a set amount of fuel. However, such engines are not in production because the manufacturing of ceramic parts in the sufficient precision and durability is difficult and costly. Processing methods often result in a wide distribution of microscopic flaws which frequently play a detrimental role in the sintering process, resulting in the proliferation of cracks, and ultimate mechanical failure.

Glass Ceramics

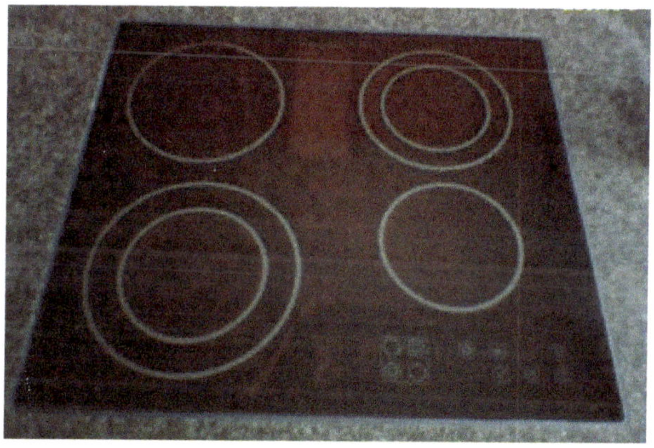

A high strength glass-ceramic cooktop with negligible thermal expansion

Glass-ceramic materials share many properties with both non-crystalline glasses and crystalline ceramics. They are formed as a glass, and then partially crystallized by heat treatment, producing both amorphous and crystalline phases so that crystalline grains are embedded within a non-crystalline intergranular phase.

Glass-ceramics are used to make cookware (originally known by the brand name CorningWare) and stovetops which have both high resistance to thermal shock and extremely low permeability to liquids. The negative coefficient of thermal expansion of the crystalline ceramic phase can be balanced with the positive coefficient of the glassy phase. At a certain point (~70% crystalline) the glass-ceramic has a net coefficient of thermal expansion close to zero. This type of glass-ceramic exhibits excellent mechanical properties and can sustain repeated and quick temperature changes up to 1000 °C.

Glass ceramics may also occur naturally when lightning strikes the crystalline (e.g. quartz) grains found in most beach sand. In this case, the extreme and immediate heat of the lightning (~2500 °C) creates hollow, branching rootlike structures called fulgurite via fusion.

Organic Solids

Organic chemistry studies the structure, properties, composition, reactions, and preparation by synthesis (or other means) of chemical compounds of carbon and hydrogen, which may contain any number of other elements such as nitrogen, oxygen and the halogens: fluorine, chlorine, bromine and iodine. Some organic compounds may also contain the elements phosphorus or sulfur. Examples of organic solids include wood, paraffin wax, naphthalene and a wide variety of polymers and plastics.

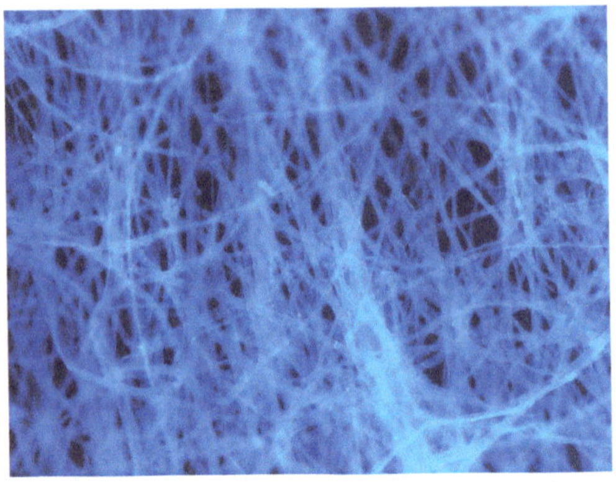

The individual wood pulp fibers in this sample are around 10 μm in diameter

Wood

Wood is a natural organic material consisting primarily of cellulose fibers embedded in a matrix of lignin. Regarding mechanical properties, the fibers are strong in tension, and the lignin matrix resists compression. Thus wood has been an important construction material since humans began building shelters and using boats. Wood to be used for construction work is commonly known as *lumber* or *timber*. In construction, wood is not only a structural material, but is also used to form the mould for concrete.

Wood-based materials are also extensively used for packaging (e.g. cardboard) and paper which are both created from the refined pulp. The chemical pulping processes use a combination of high temperature and alkaline (kraft) or acidic (sulfite) chemicals to break the chemical bonds of the lignin before burning it out.

Polymers

STM image of self-assembled supramolecular chains of the organic semiconductor quinacridone on graphite

One important property of carbon in organic chemistry is that it can form certain compounds, the individual molecules of which are capable of attaching themselves to one another, thereby forming a chain or a network. The process is called polymerization and the chains or networks polymers, while the source compound is a monomer. Two main groups of polymers exist: those artificially manufactured are referred to as industrial polymers or synthetic polymers (plastics) and those naturally occurring as biopolymers.

Monomers can have various chemical substituents, or functional groups, which can affect the chemical properties of organic compounds, such as solubility and chemical reactivity, as well as the physical properties, such as hardness, density, mechanical or tensile strength, abrasion resistance, heat resistance, transparency, color, etc.. In proteins, these differences give the polymer the ability to adopt a biologically active conformation in preference to others.

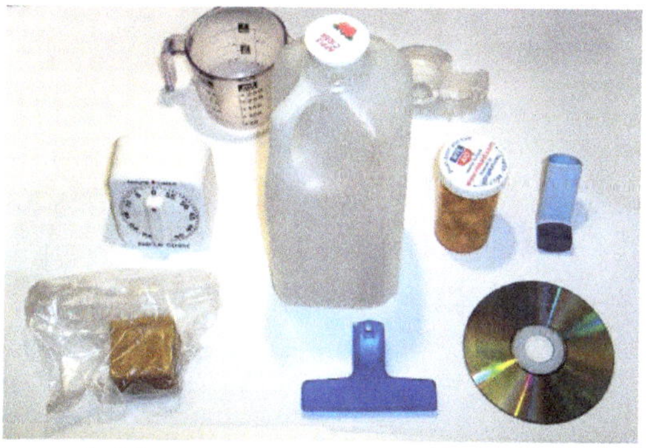

Household items made of various kinds of plastic

People have been using natural organic polymers for centuries in the form of waxes and shellac which is classified as a thermoplastic polymer. A plant polymer named cellulose provided the tensile strength for natural fibers and ropes, and by the early 19th century natural rubber was in widespread use. Polymers are the raw materials (the resins) used to make what are commonly called plastics. Plastics are the final product, created after one or more polymers or additives have been added to a resin during processing, which is then shaped into a final form. Polymers which have been around, and which are in current widespread use, include carbon-based polyethylene, polypropylene, polyvinyl chloride, polystyrene, nylons, polyesters, acrylics, polyurethane, and polycarbonates, and silicon-based silicones. Plastics are generally classified as "commodity", "specialty" and "engineering" plastics.

Composite Materials

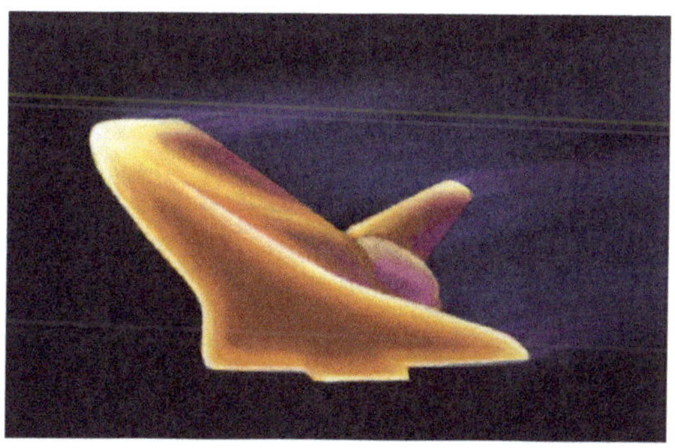

Simulation of the outside of the Space Shuttle as it heats up to over 1500 °C during re-entry

A cloth of woven carbon fiber filaments, a common element in composite materials

Composite materials contain two or more macroscopic phases, one of which is often ceramic. For example, a continuous matrix, and a dispersed phase of ceramic particles or fibers.

Applications of composite materials range from structural elements such as steel-reinforced concrete, to the thermally insulative tiles which play a key and integral role in NASA's Space Shuttle thermal protection system which is used to protect the surface of the shuttle from the heat of re-entry into the Earth's atmosphere. One example is Reinforced Carbon-Carbon (RCC), the light gray material which withstands reentry temperatures up to 1510 °C (2750 °F) and protects the nose cap and leading edges of Space Shuttle's wings. RCC is a laminated composite material made from graphite rayon cloth and impregnated with a phenolic resin. After curing at high temperature in an autoclave, the laminate is pyrolized to convert the resin to carbon, impregnated with furfural alcohol in a vacuum chamber, and cured/pyrolized to convert the furfural alcohol to carbon. In order to provide oxidation resistance for reuse capability, the outer layers of the RCC are converted to silicon carbide.

Domestic examples of composites can be seen in the "plastic" casings of television sets, cell-phones and so on. These plastic casings are usually a composite made up of a thermoplastic matrix such as acrylonitrile butadiene styrene (ABS) in which calcium carbonate chalk, talc, glass fibers or carbon fibers have been added for strength, bulk, or electro-static dispersion. These additions may be referred to as reinforcing fibers, or dispersants, depending on their purpose.

Thus, the matrix material surrounds and supports the reinforcement materials by maintaining their relative positions. The reinforcements impart their special mechanical and physical properties to enhance the matrix properties. A synergism produces material properties unavailable from the individual constituent materials, while the wide variety of matrix and strengthening materials provides the designer with the choice of an optimum combination.

Semiconductors

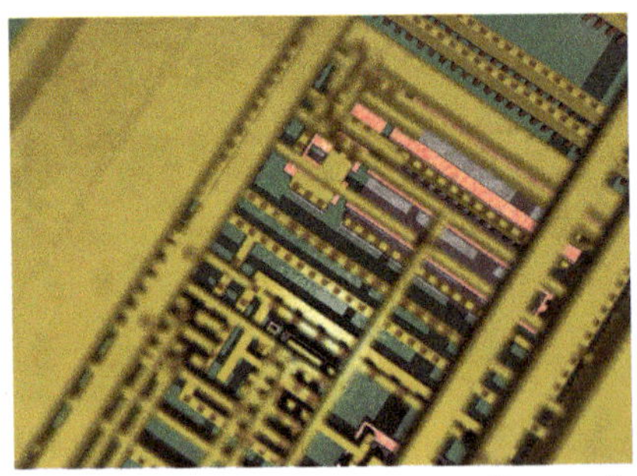

Semiconductor chip on crystalline silicon substrate

Semiconductors are materials that have an electrical resistivity (and conductivity) between that of metallic conductors and non-metallic insulators. They can be found in the periodic table moving diagonally downward right from boron. They separate the electrical conductors (or metals, to the left) from the insulators (to the right).

Devices made from semiconductor materials are the foundation of modern electronics, including radio, computers, telephones, etc. Semiconductor devices include the transistor, solar cells, diodes and integrated circuits. Solar photovoltaic panels are large semiconductor devices that directly convert light into electrical energy.

In a metallic conductor, current is carried by the flow of electrons", but in semiconductors, current can be carried either by electrons or by the positively charged "holes" in the electronic band structure of the material. Common semiconductor materials include silicon, germanium and gallium arsenide.

Nanomaterials

Bulk silicon (left) and silicon nanopowder (right)

Many traditional solids exhibit different properties when they shrink to nanometer sizes. For example, nanoparticles of usually yellow gold and gray silicon are red in color; gold nanoparticles melt at much lower temperatures (~300 °C for 2.5 nm size) than the gold slabs (1064 °C); and metallic nanowires are much stronger than the corresponding bulk metals. The high surface area of nanoparticles makes them extremely attractive for certain applications in the field of energy. For example, platinum metals may provide improvements as automotive fuel catalysts, as well as proton exchange membrane (PEM) fuel cells. Also, ceramic oxides (or cermets) of lanthanum, cerium, manganese and nickel are now being developed as solid oxide fuel cells (SOFC). Lithium, lithium–titanate and tantalum nanoparticles are being applied in lithium ion batteries. Silicon nanoparticles have been shown to dramatically expand the storage capacity of lithium ion batteries during the expansion/contraction cycle. Silicon nanowires cycle without significant degradation and present the potential for use in batteries with greatly expanded storage times. Silicon nanoparticles are also being used in new forms of solar energy cells. Thin film deposition of silicon quantum dots on the polycrystalline silicon substrate of a photovoltaic (solar) cell increases voltage output as much as 60% by fluorescing the incoming light prior to capture. Here again, surface area of the nanoparticles (and thin films) plays a critical role in maximizing the amount of absorbed radiation.

Biomaterials

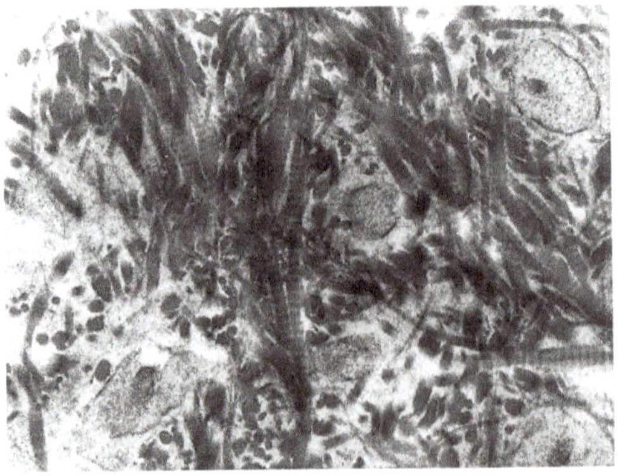

Collagen fibers of woven bone

Many natural (or biological) materials are complex composites with remarkable mechanical properties. These complex structures, which have risen from hundreds of million years of evolution, are inspiring materials scientists in the design of novel materials. Their defining characteristics include structural hierarchy, multifunctionality and self-healing capability. Self-organization is also a fundamental feature of many biological materials and the manner by which the structures are assembled from the molecular level up. Thus, self-assembly is emerging as a new strategy in the chemical synthesis of high performance biomaterials.

Physical Properties

Physical properties of elements and compounds which provide conclusive evidence of chemical composition include odor, color, volume, density (mass per unit volume), melting point, boiling point, heat capacity, physical form and shape at room temperature (solid, liquid or gas; cubic, trigonal crystals, etc.), hardness, porosity, index of refraction and many others. This section discusses some physical properties of materials in the solid state.

Mechanical

Granite rock formation in the Chilean Patagonia. Like most inorganic minerals formed by oxidation in the Earth's atmosphere, granite consists primarily of crystalline silica SiO_2 and alumina Al_2O_3

The mechanical properties of materials describe characteristics such as their strength and resistance to deformation. For example, steel beams are used in construction because of their high strength, meaning that they neither break nor bend significantly under the applied load.

Mechanical properties include elasticity and plasticity, tensile strength, compressive strength, shear strength, fracture toughness, ductility (low in brittle materials), and indentation hardness. Solid mechanics is the study of the behavior of solid matter under external actions such as external forces and temperature changes.

A solid does not exhibit macroscopic flow, as fluids do. Any degree of departure from its original shape is called deformation. The proportion of deformation to original size is called strain. If the applied stress is sufficiently low, almost all solid materials behave in such a way that the strain is directly proportional to the stress (Hooke's law). The coefficient of the proportion is called the modulus of elasticity or Young's modulus. This region of deformation is known as the linearly elastic region. Three models can describe how a solid responds to an applied stress:

- Elasticity – When an applied stress is removed, the material returns to its undeformed state.
- Viscoelasticity – These are materials that behave elastically, but also have damping. When the applied stress is removed, work has to be done against the damping effects and is converted to heat within the material. This results in a hysteresis loop in the stress–strain curve. This implies that the mechanical response has a time-dependence.
- Plasticity – Materials that behave elastically generally do so when the applied stress is less than a yield value. When the stress is greater than the yield stress, the material behaves plastically and does not return to its previous state. That is, irreversible plastic deformation (or viscous flow) occurs after yield which is permanent.

Many materials become weaker at high temperatures. Materials which retain their strength at high temperatures, called refractory materials, are useful for many purposes. For example, glass-ceramics have become extremely useful for countertop cooking, as they exhibit excellent mechanical properties and can sustain repeated and quick temperature changes up to 1000 °C. In the aerospace industry, high performance materials used in the design of aircraft and/or spacecraft exteriors must have a high resistance to thermal shock. Thus, synthetic fibers spun out of organic polymers and polymer/ceramic/metal composite materials and fiber-reinforced polymers are now being designed with this purpose in mind.

Thermal

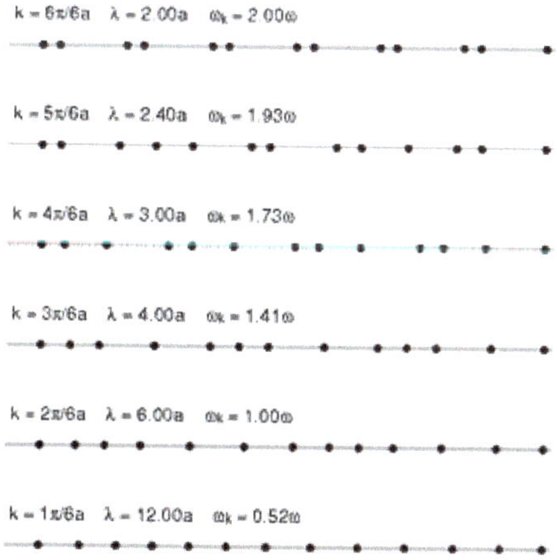

Normal modes of atomic vibration in a crystalline solid.

Because solids have thermal energy, their atoms vibrate about fixed mean positions within the ordered (or disordered) lattice. The spectrum of lattice vibrations in a crystalline or glassy network provides the foundation for the kinetic theory of solids. This motion occurs at the atomic level, and thus cannot be observed or detected without highly specialized equipment, such as that used in spectroscopy.

Thermal properties of solids include thermal conductivity, which is the property of a material that indicates its ability to conduct heat. Solids also have a specific heat capacity, which is the capacity of a material to store energy in the form of heat (or thermal lattice vibrations).

Electrical

Electrical properties include conductivity, resistance, impedance and capacitance. Electrical conductors such as metals and alloys are contrasted with electrical insulators such as glasses and ceramics. Semiconductors behave somewhere in between. Whereas conductivity in metals is caused by electrons, both electrons and holes contribute to current in semiconductors. Alternatively, ions support electric current in ionic conductors.

Many materials also exhibit superconductivity at low temperatures; they include metallic elements such as tin and aluminium, various metallic alloys, some heavily doped semiconductors, and certain ceramics. The electrical resistivity of most electrical (metallic) conductors generally decreases gradually as the temperature is lowered, but remains finite. In a superconductor however, the resistance drops abruptly to zero when the material is cooled below its critical temperature. An electric current flowing in a loop of superconducting wire can persist indefinitely with no power source.

A dielectric, or electrical insulator, is a substance that is highly resistant to the flow of electric current. A dielectric, such as plastic, tends to concentrate an applied electric field within itself which property is used in capacitors. A capacitor is an electrical device that can store energy in the electric field between a pair of closely spaced conductors (called 'plates'). When voltage is applied to the capacitor, electric charges of equal magnitude, but opposite polarity, build up on each plate. Capacitors are used in electrical circuits as energy-storage devices, as well as in electronic filters to differentiate between high-frequency and low-frequency signals.

Electro-mechanical

Piezoelectricity is the ability of crystals to generate a voltage in response to an applied mechanical stress. The piezoelectric effect is reversible in that piezoelectric crystals, when subjected to an externally applied voltage, can change shape by a small amount. Polymer materials like rubber, wool, hair, wood fiber, and silk often behave as electrets. For example, the polymer polyvinylidene fluoride (PVDF) exhibits a piezoelectric response several times larger than the traditional piezoelectric material quartz (crystalline SiO_2). The deformation (~0.1%) lends itself to useful technical applications such as high-voltage sources, loudspeakers, lasers, as well as chemical, biological, and acousto-optic sensors and/or transducers.

Optical

Materials can transmit (e.g. glass) or reflect (e.g. metals) visible light.

Many materials will transmit some wavelengths while blocking others. For example, window glass is transparent to visible light, but much less so to most of the frequencies of ultraviolet light that cause sunburn. This property is used for frequency-selective optical filters, which can alter the color of incident light.

For some purposes, both the optical and mechanical properties of a material can be of interest. For example, the sensors on an infrared homing ("heat-seeking") missile must be protected by a cover which is transparent to infrared radiation. The current material of choice for high-speed infrared-guided missile domes is single-crystal sapphire. The optical transmission of sapphire does not actually extend to cover the entire mid-infrared range (3–5 μm), but starts to drop off at wavelengths greater than approximately 4.5 μm at room temperature. While the strength of sapphire is better than that of other available mid-range infrared dome materials at room temperature, it weakens above 600 °C. A long-standing trade-off exists between optical bandpass and mechanical durability; new materials such as transparent ceramics or optical nanocomposites may provide improved performance.

Guided lightwave transmission involves the field of fiber optics and the ability of certain glasses to transmit, simultaneously and with low loss of intensity, a range of frequencies (multi-mode optical waveguides) with little interference between them. Optical waveguides are used as components in integrated optical circuits or as the transmission medium in optical communication systems.

Opto-electronic

A solar cell or photovoltaic cell is a device that converts light energy into electrical energy. Fundamentally, the device needs to fulfill only two functions: photo-generation of charge carriers (electrons and holes) in a light-absorbing material, and separation of the charge carriers to a conductive contact that will transmit the electricity (simply put, carrying electrons off through a metal contact into an external circuit). This conversion is called the photoelectric effect, and the field of research related to solar cells is known as photovoltaics.

Solar cells have many applications. They have long been used in situations where electrical power from the grid is unavailable, such as in remote area power systems, Earth-orbiting satellites and space probes, handheld calculators, wrist watches, remote radiotelephones and water pumping applications. More recently, they are starting to be used in assemblies of solar modules (photovoltaic arrays) connected to the electricity grid through an inverter, that is not to act as a sole supply but as an additional electricity source.

All solar cells require a light absorbing material contained within the cell structure to absorb photons and generate electrons via the photovoltaic effect. The materials used in solar cells tend to have the property of preferentially absorbing the wavelengths of solar light that reach the earth surface. However, some solar cells are optimized for light absorption beyond Earth's atmosphere as well.

Liquid

A liquid is a nearly incompressible fluid that conforms to the shape of its container but retains a (nearly) constant volume independent of pressure. As such, it is one of the four fundamental states of matter (the others being solid, gas, and plasma), and is the only state with a definite volume but no fixed shape. A liquid is made up of tiny vibrating particles of matter, such as atoms, held together by intermolecular bonds. Water is, by far, the most common liquid on Earth. Like a gas, a liquid is able to flow and take the shape of a container. Most liquids resist compression, although others can be compressed. Unlike a gas, a liquid does not disperse to fill every space of a container, and maintains a fairly constant density. A distinctive property of the liquid state is surface tension, leading to wetting phenomena.

The formation of a spherical droplet of liquid water minimizes the surface area, which is the natural result of surface tension in liquids.

The density of a liquid is usually close to that of a solid, and much higher than in a gas. Therefore, liquid and solid are both termed condensed matter. On the other hand, as liquids and gases share the ability to flow, they are both called fluids. Although liquid water is abundant on Earth, this state of matter is actually the least common in the known universe, because liquids require a relatively narrow temperature/pressure range to exist. Most known matter in the universe is in gaseous form (with traces of detectable solid matter) as interstellar clouds or in plasma form within stars.

Introduction

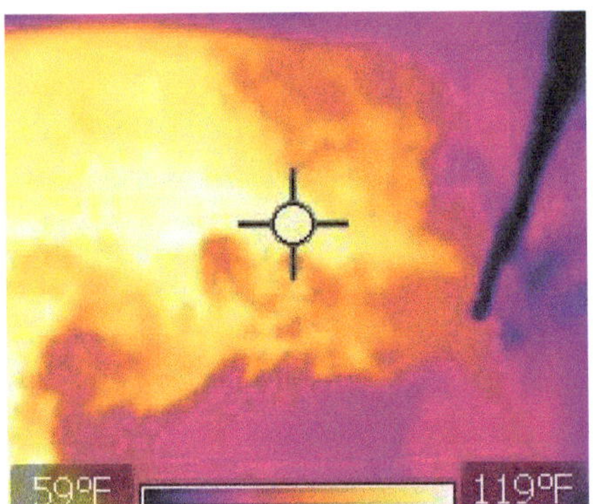

Thermal image of a sink full of hot water with cold water being added, showing how the hot and the cold water flow into each other.

Liquid is one of the four primary states of matter, with the others being solid, gas and plasma. A liquid is a fluid. Unlike a solid, the molecules in a liquid have a much greater freedom to move. The forces that bind the molecules together in a solid are only temporary in a liquid, allowing a liquid to flow while a solid remains rigid.

A liquid, like a gas, displays the properties of a fluid. A liquid can flow, assume the shape of a container, and, if placed in a sealed container, will distribute applied pressure evenly to every surface in the container. If liquid is placed in a bag, it can be squeezed into any shape. Unlike a gas, a liquid may not always mix readily with another liquid, will not always fill every space in the container, forming its own surface, and will not compress significantly, except under extremely high pressures. These properties make a liquid suitable for applications such as hydraulics.

Liquid particles are bound firmly but not rigidly. They are able to move around one another freely, resulting in a limited degree of particle mobility. As the temperature increases, the increased vibrations of the molecules causes distances between the molecules to increase. When a liquid reaches its boiling point, the cohesive forces that bind the molecules closely together break, and the liquid changes to its gaseous state (unless superheating occurs). If the temperature is decreased, the distances between the molecules become smaller. When the liquid reaches its freezing point the molecules will usually lock into a very specific order, called crystallizing, and the bonds between them become more rigid, changing the liquid into its solid state (unless supercooling occurs).

Examples

Only two elements are liquid at standard conditions for temperature and pressure: mercury and bromine. Four more elements have melting points slightly above room temperature: francium, caesium, gallium and rubidium. Metal alloys that are liquid at room temperature include NaK, a sodium-potassium metal alloy, galinstan, a fusible alloy liquid, and some amalgams (alloys involving mercury).

Pure substances that are liquid under normal conditions include water, ethanol and many other organic solvents. Liquid water is of vital importance in chemistry and biology; it is believed to be a necessity for the existence of life.

Inorganic liquids include water, magma, inorganic nonaqueous solvents and many acids.

Important everyday liquids include aqueous solutions like household bleach, other mixtures of different substances such as mineral oil and gasoline, emulsions like vinaigrette or mayonnaise, suspensions like blood, and colloids like paint and milk.

Many gases can be liquefied by cooling, producing liquids such as liquid oxygen, liquid nitrogen, liquid hydrogen and liquid helium. Not all gases can be liquified at atmospheric pressure, for example carbon dioxide can only be liquified at pressures above 5.1 atm.

Some materials cannot be classified within the classical three states of matter; they possess solid-like and liquid-like properties. Examples include liquid crystals, used in LCD displays, and biological membranes.

Applications

Liquids have a variety of uses, as lubricants, solvents, and coolants. In hydraulic systems, liquid is used to transmit power.

In tribology, liquids are studied for their properties as lubricants. Lubricants such as oil are chosen for viscosity and flow characteristics that are suitable throughout the operating temperature range of the component. Oils are often used in engines, gear boxes, metalworking, and hydraulic systems for their good lubrication properties.

Many liquids are used as solvents, to dissolve other liquids or solids. Solutions are found in a wide variety of applications, including paints, sealants, and adhesives. Naphtha and acetone are used frequently in industry to clean oil, grease, and tar from parts and machinery. Body fluids are water based solutions.

Surfactants are commonly found in soaps and detergents. Solvents like alcohol are often used as antimicrobials. They are found in cosmetics, inks, and liquid dye lasers. They are used in the food industry, in processes such as the extraction of vegetable oil.

Liquids tend to have better thermal conductivity than gases, and the ability to flow makes a liquid suitable for removing excess heat from mechanical components. The heat can be removed by channeling the liquid through a heat exchanger, such as a radiator, or the heat can be removed with the liquid during evaporation. Water or glycol coolants are used to keep engines from overheating. The coolants used in nuclear reactors include water or liquid metals, such as sodium or bismuth. Liquid propellant films are used to cool the thrust chambers of rockets. In machining, water and oils are used to remove the excess heat generated, which can quickly ruin both the work piece and the tooling. During perspiration, sweat removes heat from the human body by evaporating. In the heating, ventilation, and air-conditioning industry (HVAC), liquids such as water are used to transfer heat from one area to another.

Liquid is the primary component of hydraulic systems, which take advantage of Pascal's law to provide fluid power. Devices such as pumps and waterwheels have been used to change liquid motion into mechanical work since ancient times. Oils are forced through hydraulic pumps, which transmit this force to hydraulic cylinders. Hydraulics can be found in many applications, such as automotive brakes and transmissions, heavy equipment, and airplane control systems. Various hydraulic presses are used extensively in repair and manufacturing, for lifting, pressing, clamping and forming.

Liquids are sometimes used in measuring devices. A thermometer often uses the thermal expansion of liquids, such as mercury, combined with their ability to flow to indicate temperature. A manometer uses the weight of the liquid to indicate air pressure.

Mechanical Properties

Volume

Quantities of liquids are measured in units of volume. These include the SI unit cubic metre (m^3) and its divisions, in particular the cubic decimeter, more commonly called the litre ($1\ dm^3 = 1\ L = 0.001\ m^3$), and the cubic centimetre, also called millilitre ($1\ cm^3 = 1\ mL = 0.001\ L = 10^{-6}\ m^3$).

The volume of a quantity of liquid is fixed by its temperature and pressure. Liquids generally expand when heated, and contract when cooled. Water between 0 °C and 4 °C is a notable exception. Liquids have little compressibility. Water, for example, will compress by only 46.4 parts per million for every unit increase in atmospheric pressure (bar). At around 4000 bar (58,000 psi) of pressure, at room temperature, water only experiences an 11% decrease in volume. In the study of fluid dynamics, liquids are often treated as incompressible, especially when studying incompressible flow. This incompressible nature makes a liquid suitable for transmitting hydraulic power, because very little of the energy is lost in the form of compression. However, the very slight compressibility does lead to other phenomena. The banging of pipes, called water hammer, occurs when a valve is suddenly closed, creating a huge pressure-spike at the valve that travels backward through the system at just under the speed of sound. Another phenomenon caused by liquid's incompressibility is cavitation. Because liquids have little elasticity they can literally be pulled apart in areas of high turbulence or a dramatic change in direction, such as the trailing edge of a boat propeller or a sharp corner in a pipe. A liquid in an area of low pressure (vacuum) vaporizes and forms bubbles, which then collapse as they enter high pressure areas. This causes liquid to fill the cavities left by the bubbles with tremendous localized force, eroding any adjacent solid surface.

Pressure and Buoyancy

In a gravitational field, liquids exert pressure on the sides of a container as well as on anything within the liquid itself. This pressure is transmitted in all directions and increases with depth. If a liquid is at rest in a uniform gravitational field, the pressure, p, at any depth, z, is given by

$$p = \rho g z$$

where:

ρ is the density of the liquid (assumed constant)

g is the gravitational acceleration.

Note that this formula assumes that the pressure *at* the free surface is zero, and that surface tension effects may be neglected.

Objects immersed in liquids are subject to the phenomenon of buoyancy (Buoyancy is also observed in other fluids, but is especially strong in liquids due to their high density).

Surfaces

Surface waves in water

Unless the volume of a liquid exactly matches the volume of its container, one or more surfaces are observed. The surface of a liquid behaves like an elastic membrane in which surface tension appears, allowing the formation of drops and bubbles. Surface waves, capillary action, wetting, and ripples are other consequences of surface tension. In a confined liquid, defined by geometric constraints on a nanoscopic scale, most molecules sense some surface effects, which can result in physical properties grossly deviating from those of the bulk liquid.

Free surface

A free surface is the surface of a fluid that is subject to both zero perpendicular normal stress and parallel shear stress, such as the boundary between, e.g., liquid water and the air in the Earth's atmosphere.

Level

The liquid level (as in, e.g., water level) is the height associated with the liquid free surface, especially when it's the top-most surface. It may be measured with a level sensor.

Flow

A simulation of viscosity. The fluid on the left has a lower viscosity than the one on the right.

Viscosity measures the resistance of a liquid which is being deformed by either shear stress or extensional stress. In other words, viscosity is the resistance of a liquid to flow.

When a liquid is supercooled towards the glass transition, the viscosity increases dramatically. The liquid then becomes a viscoelastic medium that shows both the elasticity of a solid and the fluidity of a liquid, depending on the time scale of observation or on the frequency of perturbation.

Sound Propagation

The speed of sound in a fluid is given by $c = \sqrt{K / \rho}$ where K is the bulk modulus of the fluid, and ρ the density. To give a typical value, in fresh water c=1497 m/s at 25 °C.

Thermodynamics

Phase Transitions

At a temperature below the boiling point, any matter in liquid form will evaporate until the condensation of gas above reach an equilibrium. At this point the gas will condense at the same rate as the liquid evaporates. Thus, a liquid cannot exist permanently if the evaporated liquid is continually removed. A liquid at its boiling point will evaporate more quickly than the gas can condense at the current pressure. A liquid at or above its boiling point will normally boil, though superheating can prevent this in certain circumstances.

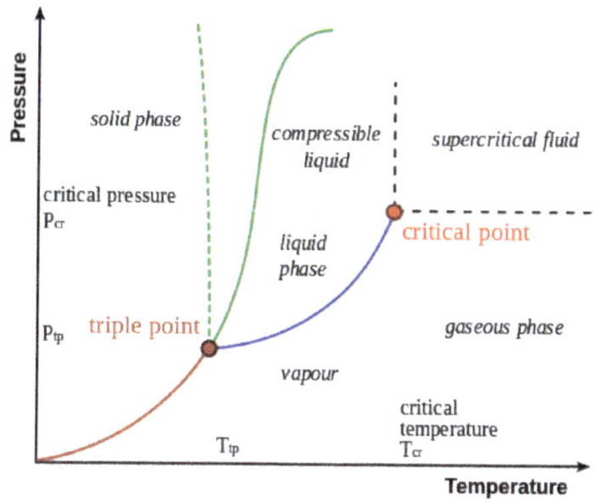

A typical phase diagram. The dotted line gives the anomalous behaviour of water. The green lines show how the freezing point can vary with pressure, and the blue line shows how the boiling point can vary with pressure. The red line shows the boundary where sublimation or deposition can occur.

At a temperature below the freezing point, a liquid will tend to crystallize, changing to its solid form. Unlike the transition to gas, there is no equilibrium at this transition under constant pressure, so unless supercooling occurs, the liquid will eventually completely crystallize. Note that this is only true under constant pressure, so e.g. water and ice in a closed, strong container might reach an equilibrium where both phases coexist.

Liquids in Space

The phase diagram explains why liquids do not exist in space or any other vacuum. Since the pressure is zero (except on surfaces or interiors of planets and moons) water and other liquids exposed to space will either immediately boil or freeze depending on the temperature. In regions of space near the earth, water will freeze if the sun is not shining directly on it and vapourize (sublime) as soon as it is in sunlight. If water exists as ice on the moon, it can only exist in shadowed holes where the sun never shines and where the surrounding rock doesn't heat it up too much. At some point near the orbit of Saturn, the light from the sun is too faint to sublime ice to water vapour. This is evident from the longevity of the ice that composes Saturn's rings.

Solutions

Liquids can display immiscibility. The most familiar mixture of two immiscible liquids in everyday life is the vegetable oil and water in Italian salad dressing. A familiar set of miscible liquids is water and alcohol. Liquid components in a mixture can often be separated from one another via fractional distillation.

Microscopic Properties

Static Structure Factor

In a liquid, atoms do not form a crystalline lattice, nor do they show any other form of long-range order. This is evidenced by the absence of Bragg peaks in X-ray and neutron diffraction. Under normal conditions, the diffraction pattern has circular symmetry, expressing the isotropy of the liquid. In radial direction, the diffraction intensity smoothly oscillates. This is usually described by the static structure factor $S(q)$, with wavenumber $q=(4\pi/\lambda)\sin\theta$ given by the wavelength λ of the probe (photon or neutron) and the Bragg angle θ. The oscillations of *S(q)* express the *near order* of the liquid, i.e. the correlations between an atom and a few shells of nearest, second nearest, ... neighbors.

A more intuitive description of these correlations is given by the radial distribution function *g(r)*, which is basically the Fourier transform of *S(q)*. It represents a spatial average of a temporal snapshot of pair correlations in the liquid.

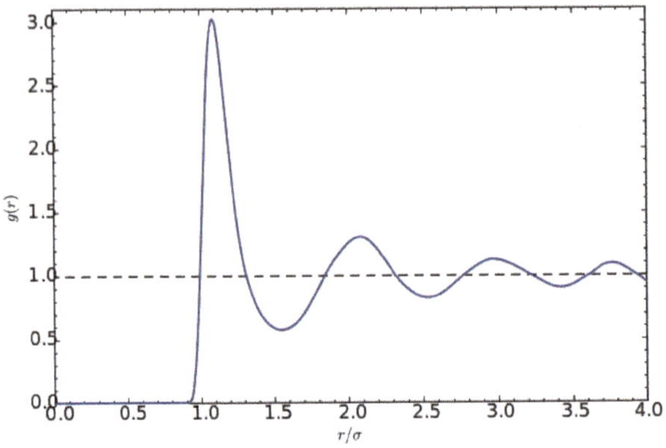

Radial distribution function of the Lennard-Jones model fluid

Sound Dispersion and Structural Relaxation

The above expression for the sound velocity $c=\sqrt{K/\rho}$ contains the bulk modulus K. If K is frequency independent then the liquid behaves as a linear medium, so that sound propagates without dissipation and without mode coupling. In reality, any liquid shows some dispersion: with increasing frequency, K crosses over from the low-frequency, liquid-like limit K_0 to the high-frequency, solid-like limit K_∞. In normal liquids, most of this cross over takes place at frequencies between GHz and THz, sometimes called hypersound.

At sub-GHz frequencies, a normal liquid cannot sustain shear waves: the zero-frequency limit of the shear modulus is $G_0=0$. This is sometimes seen as the defining property of a liquid. However, just as the bulk modulus K, the shear modulus G is frequency dependent, and at hypersound frequencies it shows a similar cross over from the liquid-like limit G_0 to a solid-like, non-zero limit G_∞.

According to the Kramers-Kronig relation, the dispersion in the sound velocity (given by the real part of *K* or *G*) goes along with a maximum in the sound attenuation (dissipation, given by the imaginary part of *K* or *G*). According to linear response theory, the Fourier transform of *K* or *G* describes how the system returns to equilibrium after an external perturbation; for this reason, the dispersion step in the GHz..THz region is also called structural relaxation. According to the fluctuation-dissipation theorem, relaxation *towards* equilibrium is intimately connected to fluctuations *in* equilibrium. The density fluctuations associated with sound waves can be experimentally observed by Brillouin scattering.

On supercooling a liquid towards the glass transition, the crossover from liquid-like to solid-like response moves from GHz to MHz, kHz, Hz, ...; equivalently, the characteristic time of structural relaxation increases from ns to μs, ms, s, ... This is the microscopic explanation for the above-mentioned viscoelastic behaviour of glass-forming liquids.

Effects of Association

The mechanisms of atomic/molecular diffusion (or particle displacement) in solids are closely related to the mechanisms of viscous flow and solidification in liquid materials. Descriptions of viscosity in terms of molecular "free space" within the liquid were modified as needed in order to account for liquids whose molecules are known to be "associated" in the liquid state at ordinary temperatures. When various molecules combine together to form an associated molecule, they enclose within a semi-rigid system a certain amount of space which before was available as free space for mobile molecules. Thus, increase in viscosity upon cooling due to the tendency of most substances to become *associated* on cooling.

Similar arguments could be used to describe the effects of pressure on viscosity, where it may be assumed that the viscosity is chiefly a function of the volume for liquids with a finite compressibility. An increasing viscosity with rise of pressure is therefore expected. In addition, if the volume is expanded by heat but reduced again by pressure, the viscosity remains the same.

The local tendency to orientation of molecules in small groups lends the liquid (as referred to previously) a certain degree of association. This association results in a considerable "internal pressure" within a liquid, which is due almost entirely to those molecules which, on account of their temporary low velocities (following the Maxwell distribution) have coalesced with other molecules. The internal pressure between several such molecules might correspond to that between a group of molecules in the solid form.

Gas

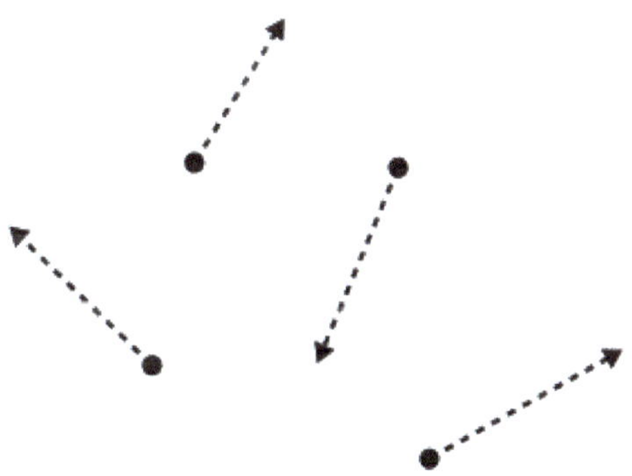

Gas phase particles (atoms, molecules, or ions) move around freely in the absence of an applied electric field

Gas is one of the four fundamental states of matter (the others being solid, liquid, and plasma). A pure gas may be made up of individual atoms (e.g. a noble gas like neon), elemental molecules made from one type of atom (e.g. oxygen), or compound molecules made from a variety of atoms (e.g. carbon dioxide). A gas mixture would contain a variety of pure gases much like the air. What distinguishes a gas from liquids and solids is the vast separation of the individual gas particles. This separation usually makes a colorless gas invisible to the human observer. The interaction of gas particles in the presence of electric and gravitational fields are considered negligible as indicated by the constant velocity vectors in the image. One type of commonly known gas is steam.

The gaseous state of matter is found between the liquid and plasma states, the latter of which provides the upper temperature boundary for gases. Bounding the lower end of the temperature scale lie degenerative quantum gases which are gaining increasing attention. High-density atomic gases super cooled to incredibly low temperatures are classified by their statistical behavior as either a Bose gas or a Fermi gas.

Elemental Gases

The only chemical elements that are stable diatomic homonuclear molecules at STP are hydrogen (H_2), nitrogen (N_2), oxygen (O_2), and two halogens: fluorine (F_2) and chlorine (Cl_2). When grouped together with the monatomic noble gases – helium (He), neon (Ne), argon (Ar), krypton (Kr), xenon (Xe), and radon (Rn) – these gases are called "elemental gases". Alternatively, they are sometimes known as "molecular gases" to distinguish them from molecules that are also chemical compounds.

Etymology

The word *gas* is a neologism first used by the early 17th-century Flemish chemist J.B. van Helmont. Van Helmont's word appears to have been simply a phonetic transcription of the Greek word *Chaos* – the *g* in Dutch being pronounced like *ch* in "loch" – in which case Van Helmont was simply following the established alchemical usage first attested in the works of Paracelsus. According to Paracelsus's terminology, *chaos* meant something like "ultra-rarefied water".

An alternative story is that Van Helmont's word is corrupted from *gahst* (or *geist*), signifying a ghost or spirit. This was because certain gases suggested a supernatural origin, such as from their ability to cause death, extinguish flames, and to occur in "mines, bottom of wells, churchyards and other lonely places".

Physical Characteristics

Drifting smoke particles provide clues to the movement of the surrounding gas

Because most gases are difficult to observe directly, they are described through the use of four physical properties or macroscopic characteristics: pressure, volume, number of particles (chemists group them by moles) and temperature. These four characteristics were repeatedly observed by scientists such as Robert Boyle, Jacques Charles, John Dalton, Joseph Gay-Lussac and Amedeo Avogadro for a variety of gases in various settings. Their detailed studies ultimately led to a mathematical relationship among these properties expressed by the ideal gas law.

Gas particles are widely separated from one another, and consequently, have weaker intermolecular bonds than liquids or solids. These intermolecular forces result from electrostatic interactions between gas particles. Like-charged areas of different gas particles repel, while oppositely charged regions of different gas particles attract one another; gases that contain permanently charged ions are known as plasmas. Gaseous compounds with polar covalent bonds contain permanent charge imbalances and so experience relatively strong intermolecular forces, although the molecule while the compound's net charge remains neutral. Transient, randomly induced charges exist across non-polar covalent bonds of molecules and electrostatic interactions caused by them are referred to as Van der Waals forces. The interaction of these intermolecular forces varies within a substance which determines many of the physical properties unique to each gas. A comparison of *boiling points* for compounds formed by ionic and covalent bonds leads us to this conclusion. The drifting smoke particles in the image provides some insight into low-pressure gas behavior.

Compared to the other states of matter, gases have low density and viscosity. Pressure and temperature influence the particles within a certain volume. This variation in particle separation and speed is referred to as *compressibility*. Finally, gas particles spread apart or diffuse in order to homogeneously distribute themselves throughout any container.

Macroscopic

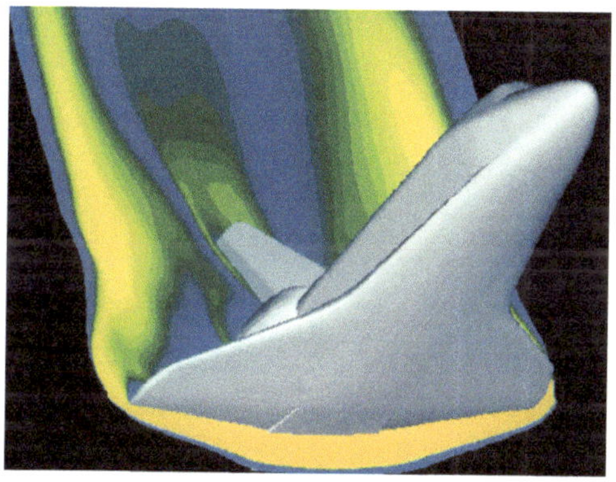

Shuttle imagery of re-entry phase

When observing a gas, it is typical to specify a frame of reference or length scale. A *larger* length scale corresponds to a macroscopic or global point of view of the gas. This region (referred to as a volume) must be sufficient in size to contain a large sampling of gas particles. The resulting statistical analysis of this sample size produces the "average" behavior (i.e. velocity, temperature or pressure) of all the gas particles within the region. In contrast, a *smaller* length scale corresponds to a microscopic or particle point of view.

Macroscopically, the gas characteristics measured are either in terms of the gas particles themselves (velocity, pressure, or temperature) or their surroundings (volume). For example, Robert Boyle studied pneumatic chemistry for a small portion of his career. One of his experiments related the macroscopic properties of pressure and volume of a gas. His experiment used a J-tube manometer which looks like a test tube in the shape of the letter J. Boyle trapped an inert gas in the closed end of the test tube with a column of mercury, thereby making the number of particles and the temperature constant. He observed that when the pressure was increased in the gas, by adding more mercury to the column, the trapped gas' volume decreased (this is known as an inverse relationship). Furthermore, when Boyle multiplied the pressure and volume of each observation, the product was constant. This relationship held for every gas that Boyle observed leading to the law, (PV=k), named to honor his work in this field.

There are many mathematical tools available for analyzing gas properties. As gases are subjected to extreme conditions, these tools become a bit more complex, from the Euler equations for inviscid flow to the Navier–Stokes equations that fully account for viscous effects. These equations are adapted to the conditions of the gas system in question. Boyle's lab equipment allowed the use of algebra to obtain his analytical results. His results were possible because he was studying gases in relatively low pressure situations where they behaved in an "ideal" manner. These ideal relationships apply to safety calculations for a variety of flight conditions on the materials in use. The high technology equipment in use today was designed to help us safely explore the more exotic operating environments where the gases no longer behave in an "ideal" manner. This advanced math, including statistics and multivariable calculus, makes possible the solution to such complex dynamic situations as space vehicle reentry. An example is the analysis of the space shuttle reentry pictured to ensure the material properties under this loading condition are appropriate. In this flight regime, the gas is no longer behaving ideally.

Pressure

The symbol used to represent *pressure* in equations is "p" or "P" with SI units of pascals.

When describing a container of gas, the term pressure (or absolute pressure) refers to the average force per unit area that the gas exerts on the surface of the container. Within this volume, it is sometimes easier to visualize the gas particles moving in straight lines until they collide with the container. The force imparted by a gas particle into the container during this collision is the change in momentum of the particle. During a collision only the normal component of velocity changes. A particle traveling parallel to the wall does not change its momentum. Therefore, the average force on a surface must be the average change in linear momentum from all of these gas particle collisions.

Pressure is the sum of all the normal components of force exerted by the particles impacting the walls of the container divided by the surface area of the wall.

Temperature

The symbol used to represent *temperature* in equations is T with SI units of kelvins.

The speed of a gas particle is proportional to its absolute temperature. The volume of the balloon in the video shrinks when the trapped gas particles slow down with the addition of extremely cold nitrogen. The temperature of any physical system is related to the motions of the particles (molecules and atoms) which make up the [gas] system. In statistical mechanics, temperature is the measure of the average kinetic energy stored in a particle. The methods of storing this energy are dictated by the degrees of freedom of the particle itself (energy modes). Kinetic energy added (endothermic process) to gas particles by way of collisions produces linear, rotational, and vibrational motion. In contrast, a molecule in a solid can only increase its vibrational modes with the addition of heat as the lattice crystal structure prevents both linear and rotational motions. These heated gas molecules have a greater speed range which constantly varies due to constant collisions with other particles. The speed range can be described by the Maxwell–Boltzmann distribution. Use of this distribution implies ideal gases near thermodynamic equilibrium for the system of particles being considered.

Specific Volume

The symbol used to represent *specific volume* in equations is "v" with SI units of cubic meters per kilogram.

The symbol used to represent volume in equations is "V" with SI units of cubic meters.

When performing a thermodynamic analysis, it is typical to speak of intensive and extensive properties. Properties which depend on the amount of gas (either by mass or volume) are called *extensive* properties, while properties that do not depend on the amount of gas are called *intensive* properties. Specific volume is an example of an *intensive* property because it is the ratio of volume occupied by a *unit of mass* of a gas that is identical throughout a system at equilibrium. 1000 atoms a gas occupy the same space as any other 1000 atoms for any given temperature and pressure. This concept is easier to visualize for solids such as iron which are incompressible compared to gases. Since a gas fills any container in which it is placed, volume is an *extensive property*.

Density

The symbol used to represent density in equations is ρ (rho) with SI units of kilograms per cubic meter. This term is the reciprocal of specific volume.

Since gas molecules can move freely within a container, their mass is normally characterized by density. Density is the amount of mass per unit volume of a substance, or the inverse of specific volume. For gases, the density can vary over a wide range because the particles are free to move closer together when constrained by pressure or volume. This variation of density is referred to as compressibility. Like pressure and temperature, density is a state variable of a gas and the change in density during any process is governed by the laws of thermodynamics. For a static gas, the density is the same throughout the entire container. Density is therefore a scalar quantity. It can be shown by kinetic theory that the density is *inversely* proportional to the size of the container in which a fixed mass of gas is confined. In this case of a fixed mass, the density decreases as the volume increases.

Microscopic

If one could observe a gas under a powerful microscope, one would see a collection of particles (molecules, atoms, ions, electrons, etc.) without any definite shape or volume that are in more or less random motion. These neutral gas particles only change direction when they collide with another particle or with the sides of the container. In an ideal gas, these collisions are perfectly elastic. This particle or microscopic view of a gas is described by the Kinetic-molecular theory.

Kinetic Theory

Kinetic theory provides insight into the macroscopic properties of gases by considering their molecular composition and motion. Starting with the definitions of momentum and kinetic energy, one can use the conservation of momentum and geometric relationships of a cube to relate macroscopic system properties of temperature and pressure to the microscopic property of kinetic energy per molecule. The theory provides averaged values for these two properties.

The theory also explains how the gas system responds to change. For example, as a gas is heated from absolute zero, when it is (in theory) perfectly still, its internal energy (temperature) is increased. As a gas is heated, the particles speed up and its temperature rises. This results in greater numbers of collisions with the container per unit time due to the higher particle speeds associated with elevated temperatures. The pressure increases in proportion to the number of collisions per unit time.

Brownian Motion

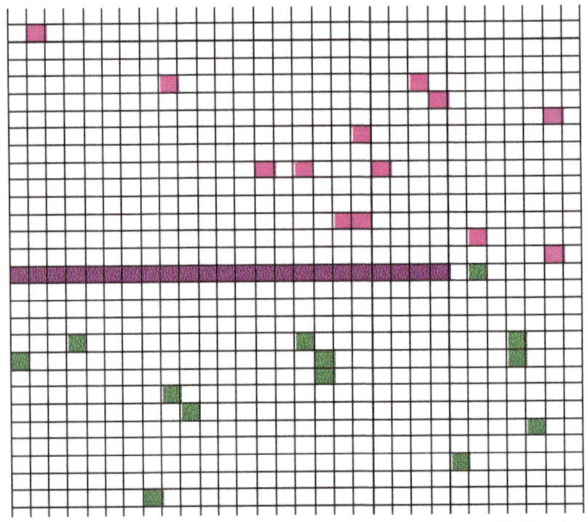

Random motion of gas particles results in diffusion

Brownian motion is the mathematical model used to describe the random movement of particles suspended in a fluid. The gas particle animation, using pink and green particles, illustrates how this behavior results in the spreading out of gases (entropy). These events are also described by particle theory.

Since it is at the limit of (or beyond) current technology to observe individual gas particles (atoms or molecules), only theoretical calculations give suggestions about how they move, but their motion is different from Brownian motion because Brownian motion involves a smooth drag due to the frictional force of many gas molecules, punctuated by violent collisions of an individual (or several) gas molecule(s) with the particle. The particle (generally consisting of millions or billions of atoms) thus moves in a jagged course, yet not so jagged as would be expected if an individual gas molecule were examined.

Intermolecular Forces

As discussed earlier, momentary attractions (or repulsions) between particles have an effect on gas dynamics. In physical chemistry, the name given to these intermolecular forces is *van der Waals force*. These forces play a key role in determining physical properties of a gas such as viscosity and flow rate. Ignoring these forces in certain conditions allows a real gas to be treated like an ideal gas. This assumption allows the use of ideal gas laws which greatly simplifies calculations.

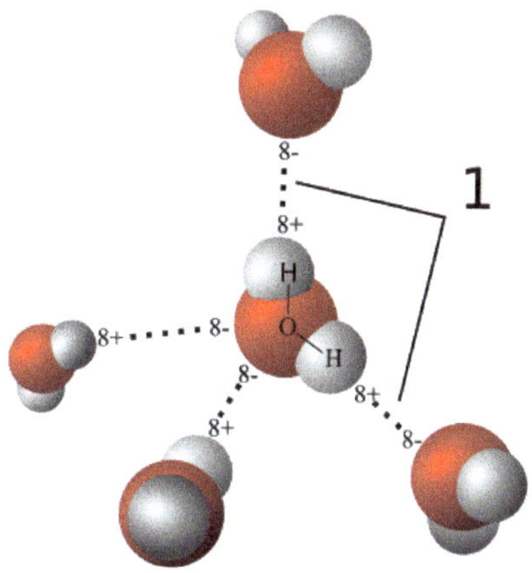

When gases are compressed, intermolecular forces like those shown here start to play a more active role

Proper use of these gas relationships requires the Kinetic-molecular theory (KMT). When gas particles possess a magnetic charge or Intermolecular force they gradually influence one another as the spacing between them is reduced (the hydrogen bond model illustrates one example). In the absence of any charge, at some point when the spacing between gas particles is greatly reduced they can no longer avoid collisions between themselves at normal gas temperatures. Another case for increased collisions among gas particles would include a fixed volume of gas, which upon heating would contain very fast particles. *This means that these ideal equations provide reasonable results* except *for extremely high pressure (compressible) or high temperature (ionized) conditions*. Notice that all of these excepted conditions allow energy transfer to take place within the gas system. The absence of these internal transfers is what is referred to as ideal conditions in which the energy exchange occurs only at the boundaries of the system. Real gases experience some of these collisions and intermolecular forces. When these collisions are statistically negligible (incompressible), results from these ideal equations are still meaningful. If the gas particles are compressed into close proximity they behave more like a liquid.

Simplified Models

An *equation of state* (for gases) is a mathematical model used to roughly describe or predict the state properties of a gas. At present, there is no single equation of state that accurately predicts the properties of all gases under all conditions. Therefore, a number of much more accurate equations of state have been developed for gases in specific temperature and pressure ranges. The "gas models" that are most widely discussed are "perfect gas", "ideal gas" and "real gas". Each of these models has its own set of assumptions to facilitate the analysis of a given thermodynamic system. Each successive model expands the temperature range of coverage to which it applies.

Ideal and Perfect Gas Models

The equation of state for an ideal or perfect gas is the ideal gas law and reads

$$PV = nRT,$$

where P is the pressure, V is the volume, n is amount of gas (in mol units), R is the universal gas constant, 8.314 J/(mol K), and T is the temperature. Written this way, it is sometimes called the "chemist's version", since it emphasizes the number of molecules n. It can also be written as

$$P = \rho R_s T,$$

where R_s is the specific gas constant for a particular gas, in units J/(kg K), and $\rho = m/V$ is density. This notation is the "gas dynamicist's" version, which is more practical in modeling of gas flows involving acceleration without chemical reactions.

The ideal gas law does not make an assumption about the specific heat of a gas. In the most general case, the specific heat is a function of both temperature and pressure. If the pressure-dependence is neglected (and possibly the temperature-dependence as well) in a particular application, sometimes the gas is said to be a perfect gas, although the exact assumptions may vary depending on the author and/or field of science.

For an ideal gas, the ideal gas law applies without restrictions on the specific heat. An ideal gas is a simplified "real gas" with the assumption that the compressibility factor Z is set to 1 meaning that this pneumatic ratio remains constant. A compressibility factor of one also requires the four state variables to follow the ideal gas law.

This approximation is more suitable for applications in engineering although simpler models can be used to produce a "ball-park" range as to where the real solution should lie. An example where the "ideal gas approximation" would be suitable would be inside a combustion chamber of a jet engine. It may also be useful to keep the elementary reactions and chemical dissociations for calculating emissions.

Real Gas

21 April 1990 eruption of Mount Redoubt, Alaska, illustrating real gases not in thermodynamic equilibrium

Each one of the assumptions listed below adds to the complexity of the problem's solution. As the density of a gas increases with rising pressure, the intermolecular forces play a more substantial role in gas behavior which results in the ideal gas law no longer providing "reasonable" results. At the upper end of the engine temperature ranges (e.g. combustor sections – 1300 K), the complex fuel particles absorb internal energy by means of rotations and vibrations that cause their specific heats to vary from those of diatomic molecules and noble gases. At more than double that temperature, electronic excitation and dissociation of the gas particles begins to occur causing the pressure to adjust to a greater number of particles (transition from gas to plasma). Finally, all of the thermodynamic processes were presumed to describe uniform gases whose velocities varied according to a fixed distribution. Using a non-equilibrium situation implies the flow field must be characterized in some manner to enable a solution. One of the first attempts to expand the boundaries of the ideal gas law was to include coverage for different thermodynamic processes by adjusting the equation to read $pV^n = constant$ and then varying the n through different values such as the specific heat ratio, γ.

Real gas effects include those adjustments made to account for a greater range of gas behavior:

- Compressibility effects (Z allowed to vary from 1.0)
- Variable heat capacity (specific heats vary with temperature)
- Van der Waals forces (related to compressibility, can substitute other equations of state)
- Non-equilibrium thermodynamic effects
- Issues with molecular dissociation and elementary reactions with variable composition.

For most applications, such a detailed analysis is excessive. Examples where "Real Gas effects" would have a significant impact would be on the Space Shuttle re-entry where extremely high temperatures and pressures are present or the gases produced during geological events as in the image of the 1990 eruption of Mount Redoubt.

Historical Synthesis

Boyle's Law

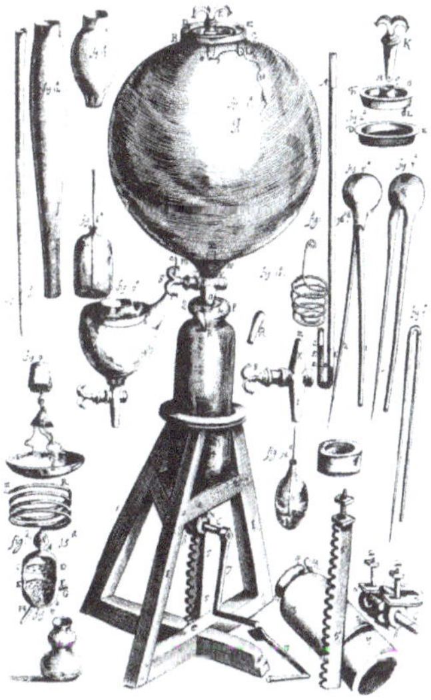

Boyle's equipment

Boyle's Law was perhaps the first expression of an equation of state. In 1662 Robert Boyle performed a series of experiments employing a J-shaped glass tube, which was sealed on one end. Mercury was added to the tube, trapping a fixed quantity of air in the short, sealed end of the tube. Then the volume of gas was carefully measured as additional mercury was added to the tube. The pressure of the gas could be determined by the difference between the mercury level in the short end of the tube and that in the long, open end. The image of Boyle's Equipment shows some of the exotic tools used by Boyle during his study of gases.

Through these experiments, Boyle noted that the pressure exerted by a gas held at a constant temperature varies inversely with the volume of the gas. For example, if the volume is halved, the pressure is doubled; and if the volume is doubled, the pressure is halved. Given the inverse relationship between pressure and volume, the product of pressure (*P*) and volume (*V*) is a constant (*k*) for a given mass of confined gas as long as the temperature is constant. Stated as a formula, thus is:

$$PV = k$$

Because the before and after volumes and pressures of the fixed amount of gas, where the before and after temperatures are the same both equal the constant *k*, they can be related by the equation:

$$P_1V_1 = P_2V_2.$$

Charles's Law

In 1787, the French physicist and balloon pioneer, Jacques Charles, found that oxygen, nitrogen, hydrogen, carbon dioxide, and air expand to the same extent over the same 80 kelvin interval. He noted that, for an ideal gas at constant pressure, the volume is directly proportional to its temperature:

$$\frac{V_1}{T_1} = \frac{V_2}{T_2}$$

Gay-Lussac's Law

In 1802, Joseph Louis Gay-Lussac published results of similar, though more extensive experiments. Gay-Lussac credited Charle's earlier work by naming the law in his honor. Gay-Lussac himself is credited with the law describing pressure, which he found in 1809. It states that the pressure exerted on a container's sides by an ideal gas is proportional to its temperature.

$$\frac{P_1}{T_1} = \frac{P_2}{T_2}$$

Avogadro's Law

In 1811, Amedeo Avogadro verified that equal volumes of pure gases contain the same number of particles. His theory was not generally accepted until 1858 when another Italian chemist Stanislao Cannizzaro was able to explain non-ideal exceptions. For his work with gases a century prior, the number that bears his name Avogadro's constant represents the number of atoms found in 12 grams of elemental carbon-12 (6.022×10^{23} mol^{-1}). This specific number of gas particles, at standard temperature and pressure (ideal gas law) occupies 22.40 liters, which is referred to as the molar volume.

Avogadro's law states that the volume occupied by an ideal gas is proportional to the number of moles (or molecules) present in the container. This gives rise to the molar volume of a gas, which at STP is 22.4 dm³ (or litres). The relation is given by

$$\frac{V_1}{n_1} = \frac{V_2}{n_2}$$

where n is equal to the number of moles of gas (the number of molecules divided by Avogadro's Number).

Dalton's Law

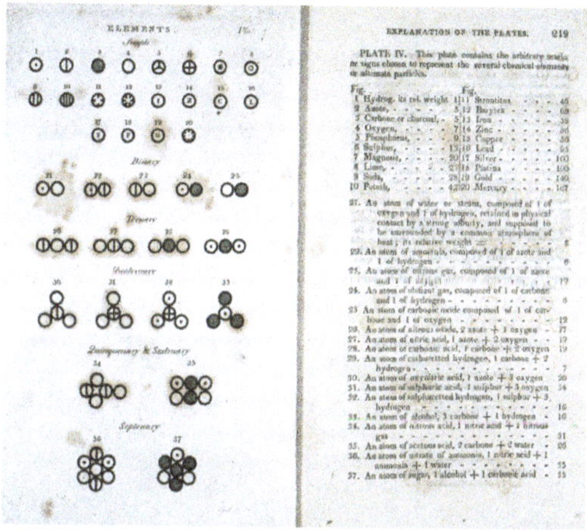

Dalton's notation

In 1801, John Dalton published the Law of Partial Pressures from his work with ideal gas law relationship: The pressure of a mixture of non reactive gases is equal to the sum of the pressures of all of the constituent gases alone. Mathematically, this can be represented for *n* species as:

$$\text{Pressure}_{\text{total}} = \text{Pressure}_1 + \text{Pressure}_2 + \ldots + \text{Pressure}_n$$

The image of Dalton's journal depicts symbology he used as shorthand to record the path he followed. Among his key journal observations upon mixing unreactive "elastic fluids" (gases) were the following:

- Unlike liquids, heavier gases did not drift to the bottom upon mixing.
- Gas particle identity played no role in determining final pressure (they behaved as if their size was negligible).

Special Topics

Compressibility

Thermodynamicists use this factor (*Z*) to alter the ideal gas equation to account for compressibility effects of real gases. This factor represents the ratio of actual to ideal specific volumes. It is sometimes referred to as a "fudge-factor" or correction to expand the useful range of the ideal gas law for design purposes. *Usually* this *Z* value is very close to unity. The compressibility factor image illustrates how Z varies over a range of very cold temperatures.

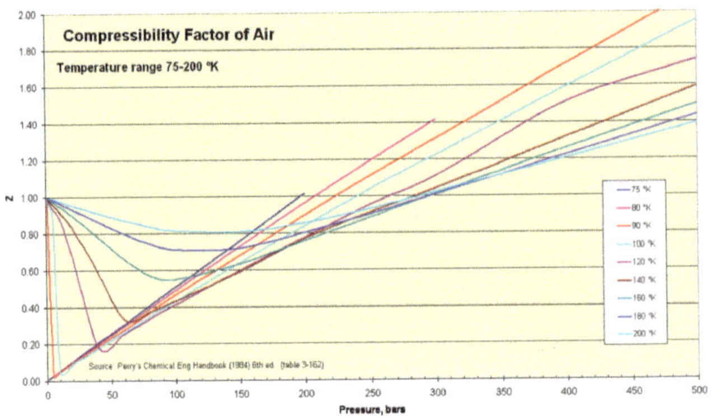

Compressibility factors for air

Reynolds Number

In fluid mechanics, the Reynolds number is the ratio of inertial forces ($v_s\rho$) to viscous forces (μ/L). It is one of the most important dimensionless numbers in fluid dynamics and is used, usually along with other dimensionless numbers, to provide a criterion for determining dynamic similitude. As such, the Reynolds number provides the link between modeling results (design) and the full-scale actual conditions. It can also be used to characterize the flow.

Viscosity

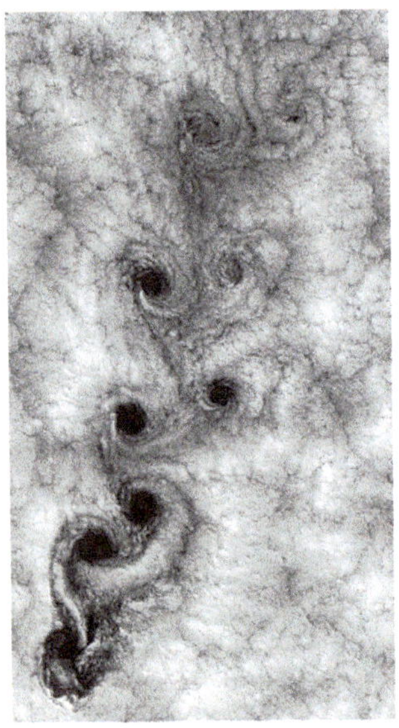

Satellite view of weather pattern in vicinity of Robinson Crusoe Islands on 15 September 1999, shows a unique turbulent cloud pattern called a Kármán vortex street.

Viscosity, a physical property, is a measure of how well adjacent molecules stick to one another. A solid can withstand a shearing force due to the strength of these sticky intermolecular forces. A fluid will continuously deform when subjected to a similar load. While a gas has a lower value of viscosity than a liquid, it is still an observable property. If gases had no viscosity, then they would not stick to the surface of a wing and form a boundary layer. A study of the delta wing in the Schlieren image reveals that the gas particles stick to one another.

Turbulence

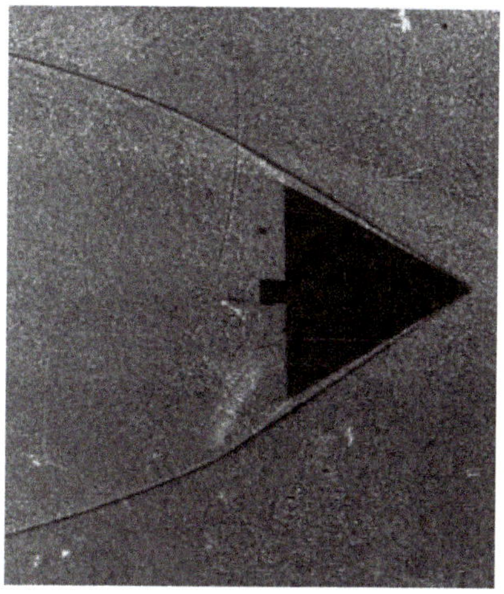

Delta wing in wind tunnel. The shadows form as the indices of refraction change within the gas as it compresses on the leading edge of this wing.

In fluid dynamics, turbulence or turbulent flow is a flow regime characterized by chaotic, stochastic property changes. This includes low momentum diffusion, high momentum convection, and rapid variation of pressure and velocity in space and time. The Satellite view of weather around Robinson Crusoe Islands illustrates just one example.

Boundary Layer

Particles will, in effect, "stick" to the surface of an object moving through it. This layer of particles is called the boundary layer. At the surface of the object, it is essentially static due to the friction of the surface. The object, with its boundary layer is effectively the new shape of the object that the rest of the molecules "see" as the object approaches. This boundary layer *can* separate from the surface, essentially creating a new surface and completely changing the flow path. The classical example of this is a stalling airfoil. The delta wing image clearly shows the boundary layer thickening as the gas flows from right to left along the leading edge.

Maximum Entropy Principle

As the total number of degrees of freedom approaches infinity, the system will be found in the macrostate that corresponds to the highest multiplicity. In order to illustrate this principle, observe the skin temperature of a frozen metal bar. Using a thermal image of the skin temperature, note the temperature distribution on the surface. This initial observation of temperature represents a "microstate." At some future time, a second observation of the skin temperature produces a second microstate. By continuing this observation process, it is possible to produce a series of microstates that illustrate the thermal history of the bar's surface. Characterization of this historical series of microstates is possible by choosing the macrostate that successfully classifies them all into a single grouping.

Thermodynamic Equilibrium

When energy transfer ceases from a system, this condition is referred to as thermodynamic equilibrium. Usually this condition implies the system and surroundings are at the same temperature so that heat no longer transfers between them. It also implies that external forces are balanced (volume does not change), and all chemical reactions within the system are complete. The timeline varies for these events depending on the system in question. A container of ice allowed to melt at room temperature takes hours, while in semiconductors the heat transfer that occurs in the device transition from an on to off state could be on the order of a few nanoseconds.

Evaporation

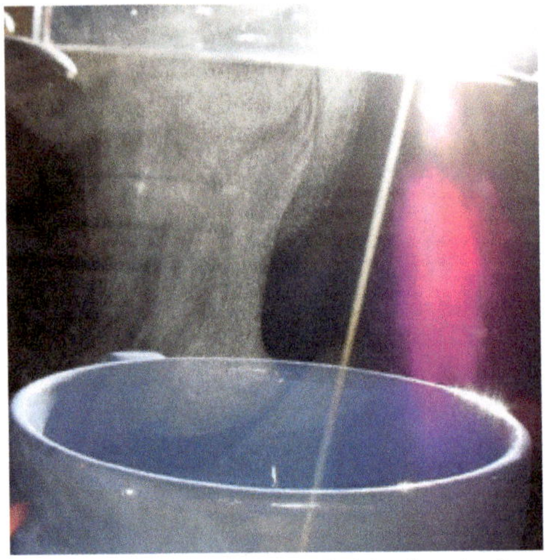

Aerosol of microscopic water droplets suspended in the air above a hot tea cup after that water vapor has sufficiently cooled and condensed. Water vapor is an invisible gas, but the clouds of condensed water droplets refract and disperse the sun light and so are visible.

Evaporation is a type of vaporization of a liquid that occurs from the surface of a liquid into a gaseous phase that is not saturated with the evaporating substance. The other type of vaporization is boiling, which is characterized by bubbles of saturated vapor forming in the liquid phase. Steam produced in a boiler is another example of evaporation occurring in a saturated vapor phase. Evaporation that occurs directly from the solid phase below the melting point, as commonly observed with ice at or below freezing or moth crystals (napthalene or paradichlorobenzene), is called sublimation.

On average, a fraction of the molecules in a glass of water have enough heat energy to escape from the liquid. The reverse also happens — water molecules from the air enter the water in the glass — but as long as the relative humidity of the air in contact is less than 100% (i.e., saturation), the net transfer of water molecules will be to the air. The water in the glass will be cooled by the evaporation until an equilibrium is reached where the air supplies the amount of heat removed by the evaporating water. In an enclosed environment the water would evaporate until the air is saturated.

With sufficient temperature, the liquid would turn into vapor quickly. When the molecules collide, they transfer energy to each other in varying degrees, based on how they collide. Sometimes the transfer is so one-sided for a molecule near the surface that it ends up with enough energy to "escape" and enter the surrounding air.

Evaporation is an essential part of the water cycle. The sun (solar energy) drives evaporation of water from oceans, lakes, moisture in the soil, and other sources of water. In hydrology, evaporation and transpiration (which involves evaporation within plant stomata) are collectively termed evapotranspiration. Evaporation of water occurs when the surface of the liquid is exposed, allowing molecules to escape and form water vapor; this vapor can then rise up and form clouds.

Theory

For molecules of a liquid to evaporate, they must be located near the surface, they have to be moving in the proper direction, and have sufficient kinetic energy to overcome liquid-phase intermolecular forces. When only a small proportion of the molecules meet these criteria, the rate of evaporation is low. Since the kinetic energy of a molecule is proportional to its temperature, evaporation proceeds more quickly at higher temperatures. As the faster-moving molecules escape, the remaining molecules have lower average kinetic energy, and the temperature of the liquid decreases. This phenomenon is also called evaporative cooling. This is why evaporating sweat cools the human body. Evaporation also tends to proceed more quickly with higher flow rates between the gaseous and liquid phase and in liquids with higher vapor pressure. For example, laundry on a clothes line will dry (by evaporation) more rapidly on a windy day than on a still day. Three key parts to evaporation are heat, atmospheric pressure (determines the percent humidity), and air movement.

On a molecular level, there is no strict boundary between the liquid state and the vapor state. Instead, there is a Knudsen layer, where the phase is undetermined. Because this layer is only a few molecules thick, at a macroscopic scale a clear phase transition interface cannot be seen.

Liquids that do not evaporate visibly at a given temperature in a given gas (e.g., cooking oil at room temperature) have molecules that do not tend to transfer energy to each other in a pattern sufficient to frequently give a molecule the heat energy necessary to turn into vapor. However, these liquids *are* evaporating. It is just that the process is much slower and thus significantly less visible.

Evaporative Equilibrium

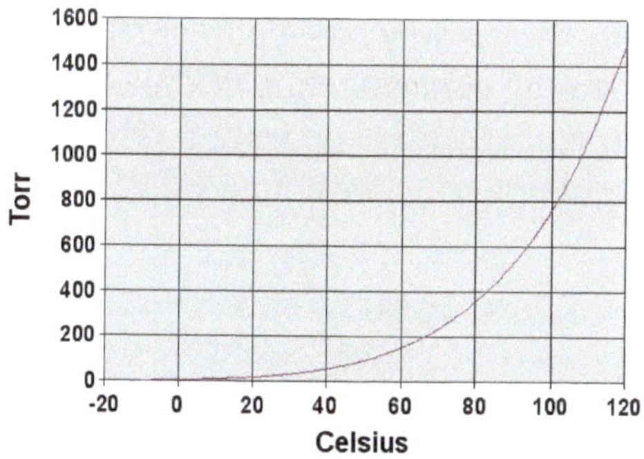

Vapor pressure of water vs. temperature. 760 Torr = 1 atm

If evaporation takes place in an enclosed area, the escaping molecules accumulate as a vapor above the liquid. Many of the molecules return to the liquid, with returning molecules becoming more frequent as the density and pressure of the vapor increases. When the process of escape and return reaches an equilibrium, the vapor is said to be "saturated", and no further change in either vapor pressure and density or liquid temperature will occur. For a system consisting of vapor and liquid of a pure substance, this equilibrium state is directly related to the vapor pressure of the substance, as given by the Clausius–Clapeyron relation:

$$\ln\left(\frac{P_2}{P_1}\right) = -\frac{\Delta H_{vap}}{R}\left(\frac{1}{T_2} - \frac{1}{T_1}\right)$$

where P_1, P_2 are the vapor pressures at temperatures T_1, T_2 respectively, ΔH_{vap} is the enthalpy of vaporization, and R is the universal gas constant. The rate of evaporation in an open system is related to the vapor pressure found in a closed system. If a liquid is heated, when the vapor pressure reaches the ambient pressure the liquid will boil.

The ability for a molecule of a liquid to evaporate is based largely on the amount of kinetic energy an individual particle may possess. Even at lower temperatures, individual molecules of a liquid can evaporate if they have more than the minimum amount of kinetic energy required for vaporization.

Factors Influencing the Rate of Evaporation

Note: Air used here is a common example; however, the vapor phase can be other gases.

Concentration of the substance evaporating in the air

> If the air already has a high concentration of the substance evaporating, then the given substance will evaporate more slowly.

Concentration of other substances in the air

> If the air is already saturated with other substances, it can have a lower capacity for the substance evaporating.

Flow rate of air

> This is in part related to the concentration points above. If "fresh" air (i.e., air which is neither already saturated with the substance nor with other substances) is moving over the substance all the time, then the concentration of the substance in the air is less likely to go up with time, thus encouraging faster evaporation. This is the result of the boundary layer at the evaporation surface decreasing with flow velocity, decreasing the diffusion distance in the stagnant layer.

Inter-molecular forces

> The stronger the forces keeping the molecules together in the liquid state, the more energy one must get to escape. This is characterized by the enthalpy of vaporization.

Pressure

> Evaporation happens faster if there is less exertion on the surface keeping the molecules from launching themselves.

Surface area

> A substance that has a larger surface area will evaporate faster, as there are more surface molecules per unit of volume that are potentially able to escape.

Temperature of the substance

> the higher the temperature of the substance the greater the kinetic energy of the molecules at its surface and therefore the faster the rate of their evaporation.

In the US, the National Weather Service measures the actual rate of evaporation from a standardized "pan" open water surface outdoors, at various locations nationwide. Others do likewise around the world. The US data is collected and compiled into an annual evaporation map. The measurements range from under 30 to over 120 inches (3,000 mm) per year.

Thermodynamics

Evaporation is an endothermic process, in that heat is absorbed during evaporation.

Applications

- Industrial applications include many printing and coating processes; recovering salts from solutions; and drying a variety of materials such as lumber, paper, cloth and chemicals.
- The use of evaporation to dry or concentrate samples is a common preparatory step for many laboratory analyses such as spectroscopy and chromatography. Systems used for this purpose include rotary evaporators and centrifugal evaporators.
- When clothes are hung on a laundry line, even though the ambient temperature is below the boiling point of water, water evaporates. This is accelerated by factors such as low humidity, heat (from the sun), and wind. In a clothes dryer, hot air is blown through the clothes, allowing water to evaporate very rapidly.
- The Matki/Matka, a traditional Indian porous clay container used for storing and cooling water and other liquids.
- The botijo, a traditional Spanish porous clay container designed to cool the contained water by evaporation.
- Evaporative coolers, which can significantly cool a building by simply blowing dry air over a filter saturated with water.

Combustion Vaporization

Fuel droplets vaporize as they receive heat by mixing with the hot gases in the combustion chamber. Heat (energy) can also be received by radiation from any hot refractory wall of the combustion chamber.

Pre-combustion Vaporization

Internal combustion engines rely upon the vaporization of the fuel in the cylinders to form a fuel/air mixture in order to burn well. The chemically correct air/fuel mixture for total burning of gasoline has been determined to be 15 parts air to one part gasoline or 15/1 by weight. Changing this to a volume ratio yields 8000 parts air to one part gasoline or 8,000/1 by volume.

Film Deposition

Thin films may be deposited by evaporating a substance and condensing it onto a substrate, or by dissolving the substance in a solvent, spreading the resulting solution thinly over a substrate, and evaporating the solvent. The Hertz–Knudsen equation is often used to estimate the rate of evaporation in these instances.

Condensation

Condensation forming in the low pressure zone above the wing of an aircraft due to adiabatic expansion

Condensation is the change of the physical state of matter from gas phase into liquid phase, and is the reverse of evaporation. The word most often refers to the water cycle. It can also be defined as the change in the state of water vapour to liquid water when in contact with a liquid or solid surface or cloud condensation nuclei within the atmosphere. When the transition happens from the gaseous phase into the solid phase directly, the change is called deposition.

Initiation

Condensation is initiated by the formation of atomic/molecular clusters of that species within its gaseous volume—like rain drop or snow flake formation within clouds—or at the contact between such gaseous phase and a liquid or solid surface.

Reversibility Scenarios

A few distinct reversibility scenarios emerge here with respect to the nature of the surface.

- absorption into the surface of a liquid (either of the same substance or one of its solvents)—is reversible as evaporation.
- adsorption (as dew droplets) onto solid surface at pressures and temperatures higher than the species' triple point—also reversible as evaporation.
- adsorption onto solid surface (as supplemental layers of solid) at pressures and temperatures lower than the species' triple point—is reversible as sublimation.

Most Common Scenarios

Condensation commonly occurs when a vapor is cooled and/or compressed to its saturation limit when the molecular density in the gas phase reaches its maximal threshold. Vapor cooling and compressing equipment that collects condensed liquids is called a "condenser".

How Condensation is Measured

Psychrometry measures the rates of condensation through evaporation into the air moisture at various atmospheric pressures and temperatures. Water is the product of its vapor condensation—condensation is the process of such phase conversion.

Applications of Condensation

Condensation is a crucial component of distillation, an important laboratory and industrial chemistry application.

Because condensation is a naturally occurring phenomenon, it can often be used to generate water in large quantities for human use. Many structures are made solely for the purpose of collecting water from condensation, such as air wells and fog fences. Such systems can often be used to retain soil moisture in areas where active desertification is occurring—so much so that some organizations educate people living in affected areas about water condensers to help them deal effectively with the situation.

It is also a crucial process in forming particle tracks in a cloud chamber. In this case, ions produced by an incident particle act as nucleation centers for the condensation of the vapor producing the visible "cloud" trails.

Furthermore, condensation is a critical step in many industrial processes, such as power generation, water desalination, thermal management, refrigeration, and air conditioning.

Biological Adaptation

Numerous living beings use water made accessible by condensation. A few examples of these are the Australian thorny devil, the darkling beetles of the Namibian coast, and the coast redwoods of the West Coast of the United States.

Condensation in Building Construction

Condensation on a window during a rain shower

Condensation in building construction is an unwanted phenomenon as it may cause dampness, mold health issues, wood rot, corrosion, weakening of mortar and masonry walls, and energy penalties due to increased heat transfer. To alleviate these issues, the indoor air humidity needs to be lowered, or air ventilation in the building needs to be improved. This can be done in a number of ways, for example opening windows, turning on extractor fans, using dehumidifiers, drying clothes outside and covering pots and pans whilst cooking. Air conditioning or ventilation systems can be installed that help remove moisture from the air, and move air throughout a building. The amount of water vapour that can be stored in the air can be increased simply by increasing the temperature. However, this can be a double edged sword as most condensation in the home occurs when warm, moisture heavy air comes into contact with a cool surface. As the air is cooled, it can no longer hold as much water vapour. This leads to deposition of water on the cool surface. This is very apparent when central heating is used in combination with single glazed windows in winter.

Interstructure condensation may be caused by thermal bridges, insufficient or lacking insulation, damp proofing or insulated glazing.

References

- David Chalmers; David Manley; Ryan Wasserman (2009). Metametaphysics: New Essays on the Foundations of Ontology. Oxford University Press. pp. 378–. ISBN 978-0-19-954604-6
- Shao, Y.; Zerda, T. W. (1998). "Phase Transitions of Liquid Crystal PAA in Confined Geometries". Journal of Physical Chemistry B. 102 (18): 3387–3394. doi:10.1021/jp9734437
- V. Bendkowsky; et al. (2009). "Observation of Ultralong-Range Rydberg Molecules". Nature. 458 (7241): 1005–8. Bibcode:2009Natur.458.1005B. PMID 19396141. doi:10.1038/nature07945
- Nahum, Alan M.; Melvin, John W. (2013-03-09). Accidental Injury: Biomechanics and Prevention. Springer Science & Business Media. ISBN 9781475722642
- J.R. Minkel (20 February 2009). "Strange but True: Superfluid Helium Can Climb Walls". Scientific American. Retrieved 23 February 2010
- Trimble, Virginia (1987). "Existence and nature of dark matter in the universe". Annual Review of Astronomy and Astrophysics. 25: 425–472. Bibcode:1987ARA&A..25..425T. doi:10.1146/annurev.aa.25.090187.002233
- Theodore Gray, The Elements: A Visual Exploration of Every Known Atom in the Universe New York: Workman Publishing, 2009 p. 127 ISBN 1-57912-814-9
- Álvarez, V. H.; Dosil, N.; Gonzalez-Cabaleiro, R.; Mattedi, S.; Martin-Pastor, M.; Iglesias, M. & Navaza, J. M.: Brønsted Ionic Liquids for Sustainable Processes: Synthesis and Physical Properties. Journal of Chemical & Engineering Data 55 (2010), Nr. 2, S. 625-632. doi: 10.1021/je900550v
- Buffat, Ph.; Borel, J.-P. (1976). "Size effect on the melting temperature of gold particles". Physical Review A. 13 (6): 2287. Bibcode:1976PhRvA..13.2287B. doi:10.1103/PhysRevA.13.2287
- Born, Max (1939). "Thermodynamics of Crystals and Melting". Journal of Chemical Physics. 7 (8): 591–604. Bibcode:1939JChPh...7..591B. doi:10.1063/1.1750497
- G.W Stewart (1930). "The Cybotactic (Molecular Group) Condition in Liquids; the Association of Molecules". Phys. Rev. 35 (7): 726. Bibcode:1930PhRv...35..726S. doi:10.1103/PhysRev.35.726
- "Modern engineering for design of liquid-propellant rocket engines" by Dieter K. Huzel, David H. Huang – American Institute of Aeronautics and Astronautics 1992 p. 99 ISBN 1-56347-013-6
- Hinshaw, Gary F. (29 January 2010). "What is the universe made of?". Universe 101. NASA website. Retrieved 17 March 2010
- Enright, Ryan (23 Jul 2014). "Dropwise Condensation on Micro- and Nanostructured Surfaces". Nanoscale and Microscale Thermophysical Engineering. 18 (3). doi:10.1080/15567265.2013.862889

2

Key Concepts of Condensed Matter Physics

Phase is a thermodynamic system within which all the physical states of matter are uniform. Su-perfluidity is a characteristic wherin a liquid has zero viscosity and hence, loses no kinetic energy when flowing. The chapter strategically encompasses and incorporates the major components and key concepts of condensed matter physics, providing a complete understanding.

Phase (Matter)

In the physical sciences, a phase is a region of space (a thermodynamic system), throughout which all physical properties of a material are essentially uniform. Examples of physical properties include density, index of refraction, magnetization and chemical composition. A simple description is that a phase is a region of material that is chemically uniform, physically distinct, and (often) mechanically separable. In a system consisting of ice and water in a glass jar, the ice cubes are one phase, the water is a second phase, and the humid air over the water is a third phase. The glass of the jar is another separate phase.

The term *phase* is sometimes used as a synonym for state of matter, but there can be several immiscible phases of the same state of matter. Also, the term *phase* is sometimes used to refer to a set of equilibrium states demarcated in terms of state variables such as pressure and temperature by a phase boundary on a phase diagram. Because phase boundaries relate to changes in the organization of matter, such as a change from liquid to solid or a more subtle change from one crystal structure to another, this latter usage is similar to the use of "phase" as a synonym for state of matter. However, the state of matter and phase diagram usages are not commensurate with the formal definition given above and the intended meaning must be determined in part from the context in which the term is used.

A small piece of rapidly melting argon ice shows the transition from solid to liquid

Types of Phases

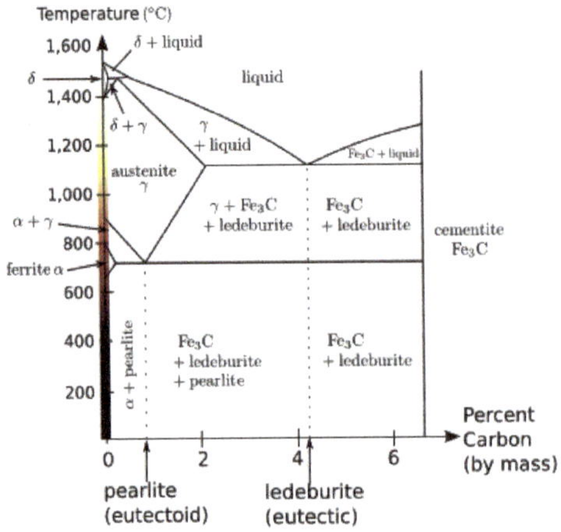

Iron-carbon phase diagram, showing the conditions necessary to form different phases

Distinct phases may be described as different states of matter such as gas, liquid, solid, plasma or Bose–Einstein condensate. Useful mesophases between solid and liquid form other states of matter.

Distinct phases may also exist within a given state of matter. As shown in the diagram for iron alloys, several phases exist for both the solid and liquid states. Phases may also be differentiated based on solubility as in polar (hydrophilic) or non-polar (hydrophobic). A mixture of water (a polar liquid) and oil (a non-polar liquid) will spontaneously separate into two phases. Water has a very low solubility (is insoluble) in oil, and oil has a low solubility in water. Solubility is the maximum amount of a solute that can dissolve in a solvent before the solute ceases to dissolve and remains in a separate phase. A mixture can separate into more than two liquid phases and the concept of phase separation extends to solids, i.e., solids can form solid solutions or crystallize into distinct crystal phases. Metal pairs that are mutually soluble can form alloys, whereas metal pairs that are mutually insoluble cannot.

As many as eight immiscible liquid phases have been observed. Mutually immiscible liquid phases are formed from water (aqueous phase), hydrophobic organic solvents, perfluorocarbons (fluorous phase), silicones, several different metals, and also from molten phosphorus. Not all organic solvents are completely miscible, e.g. a mixture of ethylene glycol and toluene may separate into two distinct organic phases.

Phases do not need to macroscopically separate spontaneously. Emulsions and colloids are examples of immiscible phase pair combinations that do not physically separate.

Phase Equilibrium

Left to equilibration, many compositions will form a uniform single phase, but depending on the temperature and pressure even a single substance may separate into two or more distinct phases. Within each phase, the properties are uniform but between the two phases properties differ.

Water in a closed jar with an air space over it forms a two phase system. Most of the water is in the liquid phase, where it is held by the mutual attraction of water molecules. Even at equilibrium molecules are constantly in motion and, once in a while, a molecule in the liquid phase gains enough kinetic energy to break away from the liquid phase and enter the gas phase. Likewise, every once in a while a vapor molecule collides with the liquid surface and condenses into the liquid. At equilibrium, evaporation and condensation processes exactly balance and there is no net change in the volume of either phase.

At room temperature and pressure, the water jar reaches equilibrium when the air over the water has a humidity of about 3%. This percentage increases as the temperature goes up. At 100 °C and atmospheric pressure, equilibrium is not reached until the air is 100% water. If the liquid is heated a little over 100 °C, the transition from liquid to gas will occur not only at the surface, but throughout the liquid volume: the water boils.

Number of Phases

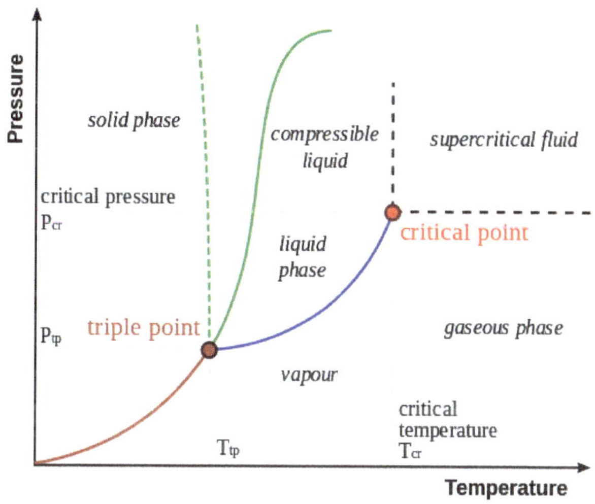

A typical phase diagram for a single-component material, exhibiting solid, liquid and gaseous phases. The solid green line shows the usual shape of the liquid–solid phase line. The dotted green line shows the anomalous behavior of water when the pressure increases. The triple point and the critical point are shown as red dots.

For a given composition, only certain phases are possible at a given temperature and pressure. The number and type of phases that will form is hard to predict and is usually determined by experiment. The results of such experiments can be plotted in phase diagrams.

The phase diagram shown here is for a single component system. In this simple system, which phases that are possible depends only on pressure and temperature. The markings show points where two or more phases can co-exist in equilibrium. At temperatures and pressures away from the markings, there will be only one phase at equilibrium.

In the diagram, the blue line marking the boundary between liquid and gas does not continue indefinitely, but terminates at a point called the critical point. As the temperature and pressure approach the critical point, the properties of the liquid and gas become progressively more similar. At the critical point, the liquid and gas become indistinguishable. Above the critical point, there are no longer separate liquid and gas phases: there is only a generic fluid phase referred to as a supercritical fluid. In water, the critical point occurs at around 647 K (374 °C or 705 °F) and 22.064 MPa.

An unusual feature of the water phase diagram is that the solid–liquid phase line (illustrated by the dotted green line) has a negative slope. For most substances, the slope is positive as exemplified by the dark green line. This unusual feature of water is related to ice having a lower density than liquid water. Increasing the pressure drives the water into the higher density phase, which causes melting.

Another interesting though not unusual feature of the phase diagram is the point where the solid–liquid phase line meets the liquid–gas phase line. The intersection is referred to as the triple point. At the triple point, all three phases can coexist.

Experimentally, the phase lines are relatively easy to map due to the interdependence of temperature and pressure that develops when multiple phases forms. Consider a test apparatus consisting of a closed and well insulated cylinder equipped with a piston. By charging the right amount of water and applying heat, the system can be brought to any point in the gas region of the phase diagram. If the piston is slowly lowered, the system will trace a curve of increasing temperature and pressure within the gas region of the phase diagram. At the point where gas begins to condense to liquid, the direction of the temperature and pressure curve will abruptly change to trace along the phase line until all of the water has condensed.

Interfacial Phenomena

Between two phases in equilibrium there is a narrow region where the properties are not that of either phase. Although this region may be very thin, it can have significant and easily observable effects, such as causing a liquid to exhibit surface tension. In mixtures, some components may preferentially move toward the interface. In terms of modeling, describing, or understanding the behavior of a particular system, it may be efficacious to treat the interfacial region as a separate phase.

Crystal Phases

A single material may have several distinct solid states capable of forming separate phases. Water is a well-known example of such a material. For example, water ice is ordinarily found in the hexagonal form ice I_h, but can also exist as the cubic ice I_c, the rhombohedral ice II, and many other forms. Polymorphism is the ability of a solid to exist in more than one crystal form. For pure chemical elements, polymorphism is known as allotropy. For example, diamond, graphite, and fullerenes are different allotropes of carbon.

Phase Transitions

When a substance undergoes a phase transition (changes from one state of matter to another) it usually either takes up or releases energy. For example, when water evaporates, the increase in kinetic energy as the evaporating molecules escape the attractive forces of the liquid is reflected in a decrease in temperature. The energy required to induce the phase transition is taken from the internal thermal energy of the water, which cools the liquid to a lower temperature; hence evaporation is useful for cooling. The reverse process, condensation, releases heat. The heat energy, or enthalpy, associated with a solid to liquid transition is the enthalpy of fusion and that associated with a solid to gas transition is the enthalpy of sublimation.

Superfluidity

Superfluidity is the characteristic property of a fluid with zero viscosity which therefore flows without loss of kinetic energy. When stirred, a superfluid forms cellular vortices that continue to rotate indefinitely. Superfluidity occurs in two isotopes of helium (helium-3 and helium-4) when they are liquified by cooling to cryogenic temperatures. It is also a property of various other exotic states of matter theorized to exist in astrophysics, high-energy physics, and theories of quantum gravity. The phenomenon is related to Bose–Einstein condensation, but neither is a specific type of the other: not all Bose-Einstein condensates can be regarded as superfluids, and not all superfluids are Bose–Einstein condensates.

Helium II will "creep" along surfaces in order to find its own level—after a short while, the levels in the two containers will equalize. The Rollin film also covers the interior of the larger container; if it were not sealed, the helium II would creep out and escape.

The liquid helium is in the superfluid phase. As long as it remains superfluid, it creeps up the wall of the cup as a thin film. It comes down on the outside, forming a drop which will fall into the liquid below. Another drop will form—and so on—until the cup is empty.

Superfluidity of Liquid Helium

Superfluidity was originally discovered in liquid helium, by Pyotr Kapitsa and John F. Allen. It has since been described through phenomenology and microscopic theories. In liquid helium-4, the superfluidity occurs at far higher temperatures than it does in helium-3. Each atom of helium-4 is a boson particle, by virtue of its integer spin. A helium-3 atom is a fermion particle; it can form bosons only by pairing with itself at much lower temperatures. The discovery of superfluidity in helium-3 was the basis for the award of the 1996 Nobel Prize in Physics. This process is similar to the electron pairing in superconductivity.

Ultracold Atomic Gases

Superfluidity in an ultracold fermionic gas was experimentally proven by Wolfgang Ketterle and his team who observed quantum vortices in ^{6}Li at a temperature of 50 nK at MIT in April 2005. Such vortices had previously been observed in an ultracold bosonic gas using ^{87}Rb in 2000, and more recently in two-dimensional gases. As early as 1999 Lene Hau created such a condensate using sodium atoms for the purpose of slowing light, and later stopping it completely. Her team then subsequently used this system of compressed light to generate the superfluid analogue of shock waves and tornadoes: "These dramatic excitations result in the formation of solitons that in turn decay into quantized vortices—created far out of equilibrium, in pairs of opposite circulation—revealing directly the process of superfluid breakdown in Bose-Einstein condensates. With a double light-roadblock setup, we can generate controlled collisions between shock waves resulting in completely unexpected, nonlinear excitations. We have observed hybrid structures consisting of vortex rings embedded in dark solitonic shells. The vortex rings act as 'phantom propellers' leading to very rich excitation dynamics."

Superfluid in Astrophysics

The idea that superfluidity exists inside neutron stars was first proposed by Arkady Migdal. By analogy with electrons inside superconductors forming Cooper pairs due to electron-lattice interaction, it is expected that nucleons in a neutron star at sufficiently high density and low temperature can also form Cooper pairs due to the long-range attractive nuclear force and lead to superfluidity and superconductivity.

Superfluidity in High-Energy Physics and Quantum Gravity

Superfluid vacuum theory (SVT) is an approach in theoretical physics and quantum mechanics where the physical vacuum is viewed as superfluid.

The ultimate goal of the approach is to develop scientific models that unify quantum mechanics (describing three of the four known fundamental interactions) with gravity. This makes SVT a candidate for the theory of quantum gravity and an extension of the Standard Model.

It is hoped that development of such theory would unify into a single consistent model of all fundamental interactions, and to describe all known interactions and elementary particles as different manifestations of the same entity, superfluid vacuum.

Electrical Resistivity and Conductivity

Electrical resistivity (also known as resistivity, specific electrical resistance, or volume resistivity) is an intrinsic property that quantifies how strongly a given material opposes the flow of electric current. A low resistivity indicates a material that readily allows the flow of electric current. Resistivity is commonly represented by the Greek letter ρ (rho). The SI unit of electrical resistivity is the ohm-metre (Ω·m). As an example, if a 1 m × 1 m × 1 m solid cube of material has sheet contacts on two opposite faces, and the resistance between these contacts is 1 Ω, then the resistivity of the material is 1 Ω·m.

Electrical conductivity or specific conductance is the reciprocal of electrical resistivity, and measures a material's ability to conduct an electric current. It is commonly represented by the Greek letter σ (sigma), but κ (kappa) (especially in electrical engineering) or γ (gamma) are also occasionally used. Its SI unit is siemens per metre (S/m) and CGSE unit is reciprocal second (s^{-1}).

Definition

Resistors or Conductors with Uniform Cross-section

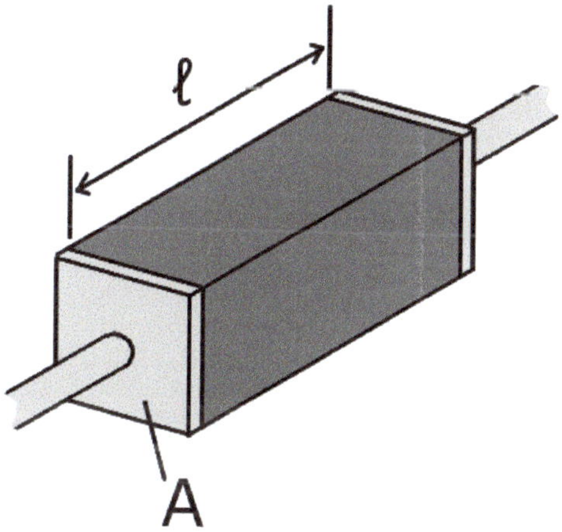

A piece of resistive material with electrical contacts on both ends

Many resistors and conductors have a uniform cross section with a uniform flow of electric current, and are made of one material. In this case, the electrical resistivity ρ (Greek: rho) is defined as:

$$\rho = R\frac{A}{\ell}.$$

where

R is the electrical resistance of a uniform specimen of the material

ℓ is the length of the piece of material

A is the cross-sectional area of the specimen

The reason resistivity is defined this way is that it makes resistivity an *intrinsic property*, unlike resistance. All copper wires, irrespective of their shape and size, have approximately the same *resistivity*, but a long, thin copper wire has a much larger *resistance* than a thick, short copper wire. Every material has its own characteristic resistivity. For example, rubber has a far larger resistivity than copper.

In a hydraulic analogy, passing current through a high-resistivity material is like pushing water through a pipe full of sand—while passing current through a low-resistivity material is like pushing water through an empty pipe. If the pipes are the same size and shape, the pipe full of sand has higher resistance to flow. Resistance, however, is not *solely* determined by the presence or absence of sand. It also depends on the length and width of the pipe: short or wide pipes have lower resistance than narrow or long pipes.

The above equation can be transposed to get Pouillet's law (named after Claude Pouillet):

$$R = \rho \frac{\ell}{A}.$$

The resistance of a given material increases with length, but decreases with increasing cross-sectional area. From the above equations, resistivity has the SI unit "ohm metre" (Ω·m).

The formula $R = \rho \ell / A$ can be used to intuitively understand the meaning of a resistivity value. For example, if A = 1 m2 ℓ = 1 m (forming a cube with perfectly conductive contacts on opposite faces), then the resistance of this element in ohms is numerically equal to the resistivity of the material it is made of in Ω·m.

Conductivity, σ, is defined as the inverse of resistivity:

$$\sigma = \frac{1}{\rho}.$$

Conductivity has SI units of "siemens per metre" (S/m).

General Definition

The above definition was specific to resistors or conductors with a uniform cross-section, where current flows uniformly through them. A more basic and general definition starts from the fact that an electric field inside a material makes electric current flow. The electrical resistivity, ρ, is defined as the ratio of the electric field to the density of the current it creates:

$$\rho = \frac{E}{J},$$

where

ρ is the resistivity of the conductor material,

E is the magnitude of the electric field,

J is the magnitude of the current density,

in which E and J are inside the conductor.

Conductivity is the inverse:

$$\sigma = \frac{1}{\rho} = \frac{J}{E}.$$

For example, rubber is a material with large ρ and small σ—because even a very large electric field in rubber makes almost no current flow through it. On the other hand, copper is a material with small ρ and large σ—because even a small electric field pulls a lot of current through it.

Causes of Conductivity

Band Theory Simplified

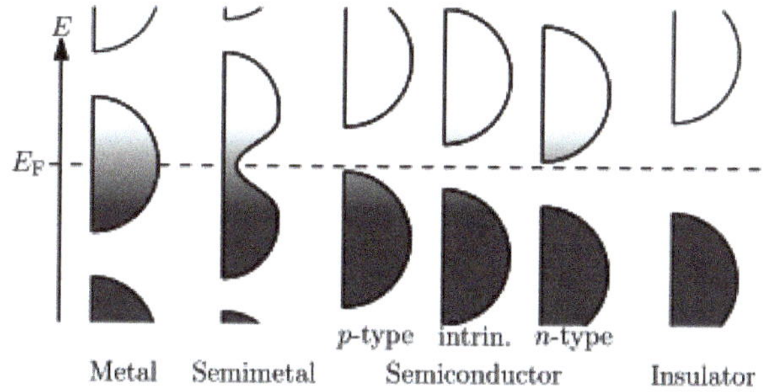

Filling of the electronic states in various types of materials at equilibrium. Here, height is energy while width is the density of available states for a certain energy in the material listed. The shade follows the Fermi–Dirac distribution (black = all states filled, white = no state filled). In metals and semimetals the Fermi level E_F lies inside at least one band. In insulators and semiconductors the Fermi level is inside a band gap; however, in semiconductors the bands are near enough to the Fermi level to be thermally populated with electrons or holes.

According to elementary quantum mechanics, electrons in an atom do not take on arbitrary energy values. Rather, electrons only occupy certain discrete energy levels in an atom or crystal; energies between these levels are impossible. When a large number of such allowed energy levels are spaced close together (in energy-space)—i.e. have similar (minutely differing) energies— we can talk about these energy levels together as an "energy band". There can be many such energy bands in a material, depending on the atomic number {number of electrons (if the atom is neutral)} and their distribution (besides external factors like environmental modification of the energy bands).

The material's electrons seek to minimize the total energy in the material by going to low energy states; however, the Pauli exclusion principle means that they cannot all go to the lowest state. The electrons instead "fill up" the band structure starting from the bottom. The characteristic energy level up to which the electrons have filled is called the Fermi level. The position of the Fermi level with respect to the band structure is very important for electrical conduction: only electrons in energy levels near the Fermi level are free to move around, since the electrons can easily jump among the partially occupied states in that region. In contrast, the low energy states are rigidly filled with a fixed number of electrons at all times, and the high energy states are empty of electrons at all times.

In metals there are many energy levels near the Fermi level, meaning that there are many electrons available to move. This is what causes the high electronic conductivity of metals.

An important part of band theory is that there may be forbidden bands in energy: energy intervals that contain no energy levels. In insulators and semiconductors, the number of electrons happens to be just the right amount to fill a certain integer number of low energy bands, exactly to the boundary. In this case, the Fermi level falls within a band gap. Since there are no available states near the Fermi level, and the electrons are not freely movable, the electronic conductivity is very low.

In Metals

Like balls in a Newton's cradle, electrons in a metal quickly transfer energy from one terminal to another, despite their own negligible movement.

A metal consists of a lattice of atoms, each with an outer shell of electrons that freely dissociate from their parent atoms and travel through the lattice. This is also known as a positive ionic lattice. This 'sea' of dissociable electrons allows the metal to conduct electric current. When an electrical potential difference (a voltage) is applied across the metal, the resulting electric field causes electrons to drift towards the positive terminal. The actual drift velocity of electrons is very small, in the order of magnitude of a meter per hour. However, as the electrons are densely packed in the material, the electromagnetic field is propagated through the metal at nearly the speed of light. The mechanism is similar to transfer of momentum of balls in a Newton's cradle.

Most metals have resistance. In simpler models (non quantum mechanical models) this can be explained by replacing electrons and the crystal lattice by a wave-like structure each. When the electron wave travels through the lattice the waves interfere, which causes resistance. The more regular the lattice is the less disturbance happens and thus resistance lowers. The amount of resistance is thus caused by mainly two factors. Firstly it is caused by the temperature and thus speed of vibration of the crystal lattice. The temperature causes irregularities in the lattice. Secondly the impurity of the metal is relevant as different ions cause irregularities too.

The larger the cross-sectional area of the conductor, the more electrons per unit length are available to carry the current. As a result, the resistance is lower in larger cross-section conductors. The number of scattering events encountered by an electron passing through a material is proportional to the length of the conductor. The longer the conductor, therefore, the higher the resistance. Different materials also affect the resistance.

In Semiconductors and Insulators

In metals, the Fermi level lies in the conduction band giving rise to free conduction electrons. However, in semiconductors the position of the Fermi level is within the band gap, approximately halfway between the conduction band minimum and valence band maximum for intrinsic (undoped) semiconductors. This means that at 0 kelvin, there are no free conduction electrons, and the resistance is infinite. However, the resistance continues to decrease as the charge carrier density in the conduction band increases. In extrinsic (doped) semiconductors, dopant atoms increase the majority charge carrier concentration by donating electrons to the conduction band or producing holes in the valence band. For both types of donor or acceptor atoms, increasing dopant density reduces resistance. Hence, highly doped semiconductors behave metallically. At very high temperatures, the contribution of thermally generated carriers dominate over the contribution from dopant atoms, and the resistance decreases exponentially with temperature.

In Ionic Liquids/Electrolytes

In electrolytes, electrical conduction happens not by band electrons or holes, but by full atomic species (ions) traveling, each carrying an electrical charge. The resistivity of ionic solutions (electrolytes) varies tremendously with concentration – while distilled water is almost an insulator, salt water is a reasonable electrical conductor. Conduction in ionic liquids is also controlled by the movement of ions, but here we are talking about molten salts rather than solvated ions. In biological membranes, currents are carried by ionic salts. Small holes in cell membranes, called ion channels, are selective to specific ions and determine the membrane resistance.

Superconductivity

The electrical resistivity of a metallic conductor decreases gradually as temperature is lowered. In ordinary conductors, such as copper or silver, this decrease is limited by impurities and other defects. Even near absolute zero, a real sample of a normal conductor shows some resistance. In a superconductor, the resistance drops abruptly to zero when the material is cooled below its critical temperature. An electric current flowing in a loop of superconducting wire can persist indefinitely with no power source.

In 1986, researchers discovered that some cuprate-perovskite ceramic materials have much higher critical temperatures, and in 1987 one was produced with a critical temperature above 90 K (−183 °C). Such a high transition temperature is theoretically impossible for a conventional superconductor, so the researchers named these conductors *high-temperature superconductors*. Liquid nitrogen boils at 77 K, facilitating many experiments and applications that are less practical at lower temperatures. In conventional superconductors, electrons are held together in pairs by an attraction mediated by lattice phonons. The best available model of high-temperature superconductivity is still somewhat crude. There is a hypothesis that electron pairing in high-temperature superconductors is mediated by short-range spin waves known as paramagnons.

Plasma

Plasmas are very good electrical conductors and electric potentials play an important role. The potential as it exists on average in the space between charged particles, independent of the question of how it can be measured, is called the *plasma potential*, or *space potential*. If an electrode is inserted into a plasma, its potential generally lies considerably below the plasma potential, due to what is termed a Debye sheath. The good electrical conductivity of plasmas makes their electric fields very small. This results in the important concept of *quasineutrality*, which says the density of negative charges is approximately equal to the density of positive charges over large volumes of the plasma ($n_e = \langle Z \rangle n_i$), but on the scale of the Debye length there can be charge imbalance. In the special case that *double layers* are formed, the charge separation can extend some tens of Debye lengths.

Lightning is an example of plasma present at Earth's surface. Typically, lightning discharges 30,000 amperes at up to 100 million volts, and emits light, radio waves, and X-rays. Plasma temperatures in lightning can approach 28,000 Kelvin (28,000 °C) (50,000 °F) and electron densities may exceed 10^{24} m^{-3}.

The magnitude of the potentials and electric fields must be determined by means other than simply finding the net charge density. A common example is to assume that the electrons satisfy the Boltzmann relation:

$$n_e \propto e^{e\Phi/k_B T_e}.$$

Differentiating this relation provides a means to calculate the electric field from the density:

$$\vec{E} = -\frac{k_B T_e}{e}\frac{\nabla n_e}{n_e}.$$

It is possible to produce a plasma that is not quasineutral. An electron beam, for example, has only negative charges. The density of a non-neutral plasma must generally be very low, or it must be very small. Otherwise, the repulsive electrostatic force dissipates it.

In astrophysical plasmas, Debye screening prevents electric fields from directly affecting the plasma over large distances, i.e., greater than the Debye length. However, the existence of charged particles causes the plasma to generate, and be affected by, magnetic fields. This can and does cause extremely complex behavior, such as the generation of plasma double layers, an object that separates charge over a few tens of Debye lengths. The dynamics of plasmas interacting with external and self-generated magnetic fields are studied in the academic discipline of magnetohydrodynamics.

Plasma is often called the *fourth state of matter* after solid, liquids and gases. It is distinct from these and other lower-energy states of matter. Although it is closely related to the gas phase in that it also has no definite form or volume, it differs in a number of ways, including the following:

Property	**Gas**	**Plasma**
Electrical conductivity	Very low: air is an excellent insulator until it breaks down into plasma at electric field strengths above 30 kilovolts per centimeter.	Usually very high: for many purposes, the conductivity of a plasma may be treated as infinite.

Independently acting species	One: all gas particles behave in a similar way, influenced by gravity and by collisions with one another.	Two or three: electrons, ions, protons and neutrons can be distinguished by the sign and value of their charge so that they behave independently in many circumstances, with different bulk velocities and temperatures, allowing phenomena such as new types of waves and instabilities.
Velocity distribution	Maxwellian: collisions usually lead to a Maxwellian velocity distribution of all gas particles, with very few relatively fast particles.	Often non-Maxwellian: collisional interactions are often weak in hot plasmas and external forcing can drive the plasma far from local equilibrium and lead to a significant population of unusually fast particles.
Interactions	Binary: two-particle collisions are the rule, three-body collisions extremely rare.	Collective: waves, or organized motion of plasma, are very important because the particles can interact at long ranges through the electric and magnetic forces.

Resistivity and Conductivity of Various Materials

- A conductor such as a metal has high conductivity and a low resistivity.
- An insulator like glass has low conductivity and a high resistivity.
- The conductivity of a semiconductor is generally intermediate, but varies widely under different conditions, such as exposure of the material to electric fields or specific frequencies of light, and, most important, with temperature and composition of the semiconductor material.

The degree of doping in semiconductors makes a large difference in conductivity. To a point, more doping leads to higher conductivity. The conductivity of a solution of water is highly dependent on its concentration of dissolved salts, and other chemical species that ionize in the solution. Electrical conductivity of water samples is used as an indicator of how salt-free, ion-free, or impurity-free the sample is; the purer the water, the lower the conductivity (the higher the resistivity). Conductivity measurements in water are often reported as *specific conductance*, relative to the conductivity of pure water at 25 °C. An EC meter is normally used to measure conductivity in a solution. A rough summary is as follows:

Material	**Resistivity, ρ (Ω·m)**
Superconductors	0
Metals	10^{-8}
Semiconductors	Variable
Electrolytes	Variable
Insulators	10^{16}
Superinsulators	∞

This table shows the resistivity, conductivity and temperature coefficient of various materials at 20 °C (68 °F, 293 K)

Material	**ρ (Ω·m) at 20 °C**	**σ (S/m) at 20 °C**	**Temperature Coefficient (K^{-1})**
Carbon (graphene)	1.00×10^{-8}	1.00×10^{8}	−0.0002
Silver	1.59×10^{-8}	6.30×10^{7}	0.0038

Copper	1.68×10^{-8}	5.96×10^{7}	0.003862
Annealed copper	1.72×10^{-8}	5.80×10^{7}	0.00393
Gold	2.44×10^{-8}	4.10×10^{7}	0.0034
Aluminium	2.82×10^{-8}	3.50×10^{7}	0.0039
Calcium	3.36×10^{-8}	2.98×10^{7}	0.0041
Tungsten	5.60×10^{-8}	1.79×10^{7}	0.0045
Zinc	5.90×10^{-8}	1.69×10^{7}	0.0037
Nickel	6.99×10^{-8}	1.43×10^{7}	0.006
Lithium	9.28×10^{-8}	1.08×10^{7}	0.006
Iron	9.71×10^{-8}	1.00×10^{7}	0.005
Platinum	1.06×10^{-7}	9.43×10^{6}	0.00392
Tin	1.09×10^{-7}	9.17×10^{6}	0.0045
Carbon steel (1010)	1.43×10^{-7}	6.99×10^{6}	
Lead	2.20×10^{-7}	4.55×10^{6}	0.0039
Titanium	4.20×10^{-7}	2.38×10^{6}	0.0038
Grain oriented electrical steel	4.60×10^{-7}	2.17×10^{6}	
Manganin	4.82×10^{-7}	2.07×10^{6}	0.000002
Constantan	4.90×10^{-7}	2.04×10^{6}	0.000008
Stainless steel	6.90×10^{-7}	1.45×10^{6}	0.00094
Mercury	9.80×10^{-7}	1.02×10^{6}	0.0009
Nichrome	1.10×10^{-6}	6.7×10^{5}	0.0004
GaAs	1.00×10^{-3} to 1.00×10^{8}	1.00×10^{-8} to 10^{3}	
Carbon (amorphous)	5.00×10^{-4} to 8.00×10^{-4}	1.25×10^{3} to 2×10^{3}	−0.0005
Carbon (graphite)	2.50×10^{-6} to 5.00×10^{-6} $\parallel$basal plane 3.00×10^{-3} $\perp$basal plane	2.00×10^{5} to 3.00×10^{5} $\parallel$basal plane 3.30×10^{2} $\perp$basal plane	
PEDOT:PSS	2×10^{-6} to 1×10^{-1}	1×10^{1} to 4.6×10^{5}	?
Germanium	4.60×10^{-1}	2.17	−0.048
Sea water	2.00×10^{-1}	4.80	
Swimming pool water	3.33×10^{-1} to 4.00×10^{-1}	0.25 to 0.30	
Drinking water	2.00×10^{1} to 2.00×10^{3}	5.00×10^{-4} to 5.00×10^{-2}	
Silicon	6.40×10^{2}	1.56×10^{-3}	−0.075
Wood (damp)	1.00×10^{3} to 1.00×10^{4}	10^{-4} to 10^{-3}	
Deionized water	1.80×10^{5}	5.50×10^{-6}	
Glass	1.00×10^{11} to 1.00×10^{15}	10^{-15} to 10^{-11}	?
Hard rubber	1.00×10^{13}	10^{-14}	?
Wood (oven dry)	1.00×10^{14} to 1.00×10^{16}	10^{-16} to 10^{-14}	
Sulfur	1.00×10^{15}	10^{-16}	?
Air	1.30×10^{14} to 3.30×10^{14}	3×10^{-15} to 8×10^{-15}	
Carbon (diamond)	1.00×10^{12}	$\sim10^{-13}$	
Fused quartz	7.50×10^{17}	1.30×10^{-18}	?
PET	1.00×10^{21}	10^{-21}	?
Teflon	1.00×10^{23} to 1.00×10^{25}	10^{-25} to 10^{-23}	?

The effective temperature coefficient varies with temperature and purity level of the material. The 20 °C value is only an approximation when used at other temperatures. For example, the coefficient becomes lower at higher temperatures for copper, and the value 0.00427 is commonly specified at 0 °C.

The extremely low resistivity (high conductivity) of silver is characteristic of metals. George Gamow tidily summed up the nature of the metals' dealings with electrons in his science-popularizing book, *One, Two, Three...Infinity* (1947):

The metallic substances differ from all other materials by the fact that the outer shells of their atoms are bound rather loosely, and often let one of their electrons go free. Thus the interior of a metal is filled up with a large number of unattached electrons that travel aimlessly around like a crowd of displaced persons. When a metal wire is subjected to electric force applied on its opposite ends, these free electrons rush in the direction of the force, thus forming what we call an electric current.

More technically, the free electron model gives a basic description of electron flow in metals.

Wood is widely regarded as an extremely good insulator, but its resistivity is sensitively dependent on moisture content, with damp wood being a factor of at least 10^{10} worse insulator than oven-dry. In any case, a sufficiently high voltage – such as that in lightning strikes or some high-tension powerlines – can lead to insulation breakdown and electrocution risk even with apparently dry wood.

Temperature Dependence

Linear Approximation

The electrical resistivity of most materials changes with temperature. If the temperature T does not vary too much, a linear approximation is typically used:

$$\rho(T) = \rho_0[1 + \alpha(T - T_0)]$$

where α is called the temperature coefficient of resistivity, T_0 is a fixed reference temperature (usually room temperature), and ρ_0 is the resistivity at temperature T_0. The parameter α is an empirical parameter fitted from measurement data. Because the linear approximation is only an approximation, α is different for different reference temperatures. For this reason it is usual to specify the temperature that α was measured at with a suffix, such as α_{15}, and the relationship only holds in a range of temperatures around the reference. When the temperature varies over a large temperature range, the linear approximation is inadequate and a more detailed analysis and understanding should be used.

Metals

In general, electrical resistivity of metals increases with temperature. Electron–phonon interactions can play a key role. At high temperatures, the resistance of a metal increases linearly with temperature. As the temperature of a metal is reduced, the temperature dependence of resistivity follows a power law function of temperature. Mathematically the temperature dependence of the resistivity ρ of a metal is given by the Bloch–Grüneisen formula:

$$\rho(T)=\rho(0)+A\left(\frac{T}{\Theta_R}\right)^n\int_0^{\frac{\Theta_R}{T}}\frac{x^n}{(e^x-1)(1-e^{-x})}dx$$

where $\rho(0)$ is the residual resistivity due to defect scattering, A is a constant that depends on the velocity of electrons at the Fermi surface, the Debye radius and the number density of electrons in the metal. Θ_R is the Debye temperature as obtained from resistivity measurements and matches very closely with the values of Debye temperature obtained from specific heat measurements. n is an integer that depends upon the nature of interaction:

1. n=5 implies that the resistance is due to scattering of electrons by phonons (as it is for simple metals)
2. n=3 implies that the resistance is due to s-d electron scattering (as is the case for transition metals)
3. n=2 implies that the resistance is due to electron–electron interaction.

If more than one source of scattering is simultaneously present, Matthiessen's Rule (first formulated by Augustus Matthiessen in the 1860s) says that the total resistance can be approximated by adding up several different terms, each with the appropriate value of *n*.

As the temperature of the metal is sufficiently reduced (so as to 'freeze' all the phonons), the resistivity usually reaches a constant value, known as the residual resistivity. This value depends not only on the type of metal, but on its purity and thermal history. The value of the residual resistivity of a metal is decided by its impurity concentration. Some materials lose all electrical resistivity at sufficiently low temperatures, due to an effect known as superconductivity.

An investigation of the low-temperature resistivity of metals was the motivation to Heike Kamerlingh Onnes's experiments that led in 1911 to discovery of superconductivity.

Semiconductors

In general, intrinsic semiconductor resistivity decreases with increasing temperature. The electrons are bumped to the conduction energy band by thermal energy, where they flow freely, and in doing so leave behind holes in the valence band, which also flow freely. The electric resistance of a typical intrinsic (non doped) semiconductor decreases exponentially with temperature:

$$\rho=\rho_0 e^{-aT}$$

An even better approximation of the temperature dependence of the resistivity of a semiconductor is given by the Steinhart–Hart equation:

$$\frac{1}{T}=A+B\ln(\rho)+C(\ln(\rho))^3$$

where *A*, *B* and *C* are the so-called Steinhart–Hart coefficients.

This equation is used to calibrate thermistors.

Extrinsic (doped) semiconductors have a far more complicated temperature profile. As temperature increases starting from absolute zero they first decrease steeply in resistance as the carriers leave the donors or acceptors. After most of the donors or acceptors have lost their carriers, the resistance starts to increase again slightly due to the reducing mobility of carriers (much as in a metal). At higher temperatures, they behave like intrinsic semiconductors as the carriers from the donors/acceptors become insignificant compared to the thermally generated carriers.

In non-crystalline semiconductors, conduction can occur by charges quantum tunnelling from one localised site to another. This is known as variable range hopping and has the characteristic form of

$$\rho = A\exp\left(T^{-\frac{1}{n}}\right),$$

where $n = 2, 3, 4$, depending on the dimensionality of the system.

Complex Resistivity and Conductivity

When analyzing the response of materials to alternating electric fields (dielectric spectroscopy), in applications such as electrical impedance tomography, it is convenient to replace resistivity with a complex quantity called impeditivity (in analogy to electrical impedance). Impeditivity is the sum of a real component, the resistivity, and an imaginary component, the reactivity (in analogy to reactance). The magnitude of impeditivity is the square root of sum of squares of magnitudes of resistivity and reactivity.

Conversely, in such cases the conductivity must be expressed as a complex number (or even as a matrix of complex numbers, in the case of anisotropic materials) called the *admittivity*. Admittivity is the sum of a real component called the conductivity and an imaginary component called the susceptivity.

An alternative description of the response to alternating currents uses a real (but frequency-dependent) conductivity, along with a real permittivity. The larger the conductivity is, the more quickly the alternating-current signal is absorbed by the material (i.e., the more opaque the material is).

Tensor Equations for Anisotropic Materials

Some materials are anisotropic, meaning they have different properties in different directions. For example, a crystal of graphite consists microscopically of a stack of sheets, and current flows very easily through each sheet, but moves much less easily from one sheet to the next.

For an anisotropic material, it is not generally valid to use the scalar equations

$$J = \sigma E \rightleftharpoons E = \rho J.$$

For example, the current may not flow in exactly the same direction as the electric field. Instead, the equations are generalized to the 3D tensor form

$$\mathbf{J} = \sigma \mathbf{E} \rightleftharpoons \mathbf{E} = \rho \mathbf{J}$$

where the conductivity σ and resistivity ρ are rank-2 tensors (in other words, 3×3 matrices). The equations are compactly illustrated in component form (using index notation and the summation convention):

$$J_i = \sigma_{ij} E_j \rightleftharpoons E_i = \rho_{ij} J_j.$$

The σ and ρ tensors are inverses (in the sense of a matrix inverse). The individual components are not necessarily inverses; for example, σ_{xx} may not be equal to $1/\rho_{xx}$.

Resistance Versus Resistivity in Complicated Geometries

Even if the material's resistivity is known, calculating the resistance of something made from it may, in some cases, be much more complicated than the formula $R = \rho\ell / A$ above. One example is spreading resistance profiling, where the material is inhomogeneous (different resistivity in different places), and the exact paths of current flow are not obvious.

In cases like this, the formulas

$$J = \sigma E \rightleftharpoons E = \rho J$$

must be replaced with

$$\mathbf{J}(\mathbf{r}) = \sigma(\mathbf{r})\mathbf{E}(\mathbf{r}) \rightleftharpoons \mathbf{E}(\mathbf{r}) = \rho(\mathbf{r})\mathbf{J}(\mathbf{r}),$$

where E and J are now vector fields. This equation, along with the continuity equation for J and the Poisson's equation for E, form a set of partial differential equations. In special cases, an exact or approximate solution to these equations can be worked out by hand, but for very accurate answers in complex cases, computer methods like finite element analysis may be required.

Resistivity Density Products

In some applications where the weight of an item is very important resistivity density products are more important than absolute low resistivity – it is often possible to make the conductor thicker to make up for a higher resistivity; and then a low resistivity density product material (or equivalently a high conductance to density ratio) is desirable. For example, for long distance overhead power lines, aluminium is frequently used rather than copper because it is lighter for the same conductance.

Material	Resistivity (nΩ·m)	Density (g/cm³)	Resistivity-density		Conductor cross-section/volume, at same conductance relative to copper
			(nΩ·m·g/cm³)	Relative to copper	
Sodium	47.7	0.97	46	31%	2.843
Lithium	92.8	0.53	49	33%	5.531
Calcium	33.6	1.55	52	35%	2.002

Material	Resistivity (nΩ·m)	Density (g/cm³)	Resistivity-density (nΩ·m·g/cm³)		Conductor cross-section/volume, at same conductance relative to copper
			(nΩ·m·g/cm³)	Relative to copper	
Potassium	72.0	0.89	64	43%	4.291
Beryllium	35.6	1.85	66	44%	2.122
Aluminium	26.50	2.70	72	48%	1.5792
Magnesium	43.90	1.74	76.3	50.9%	2.616
Copper	16.78	8.96	150	100%	1
Silver	15.87	10.49	166	111%	0.946
Gold	22.14	19.30	427	285%	1.319
Iron	96.1	7.874	757	505%	5.727

Silver, although it is the least resistive metal known, has a high density and does poorly by this measure. Calcium and the alkali metals have the best resistivity-density products, but are rarely used for conductors due to their high reactivity with water and oxygen. Aluminium is far more stable. Two other important attributes, price and toxicity, exclude the (otherwise) best choice: Beryllium. Thus, aluminium is usually the metal of choice when the weight or cost of a conductor is the driving consideration.

Spin (Physics)

In quantum mechanics and particle physics, spin is an intrinsic form of angular momentum carried by elementary particles, composite particles (hadrons), and atomic nuclei.

Spin is one of two types of angular momentum in quantum mechanics, the other being *orbital angular momentum*. The orbital angular momentum operator is the quantum-mechanical counterpart to the classical angular momentum of orbital revolution: it arises when a particle executes a rotating or twisting trajectory (such as when an electron orbits a nucleus). The existence of spin angular momentum is inferred from experiments, such as the Stern–Gerlach experiment, in which particles are observed to possess angular momentum that cannot be accounted for by orbital angular momentum alone.

In some ways, spin is like a vector quantity; it has a definite magnitude, and it has a "direction" (but quantization makes this "direction" different from the direction of an ordinary vector). All elementary particles of a given kind have the same magnitude of spin angular momentum, which is indicated by assigning the particle a *spin quantum number*.

The SI unit of spin is the (N·m·s) or ($kg{\cdot}m^2{\cdot}s^{-1}$), just as with classical angular momentum. In practice, spin is given as a dimensionless *spin quantum number* by dividing the spin angular momentum by the reduced Planck constant $\hbar$, which has the same units of angular momentum, although it should be noted that this is not the full computation of this value. Very often, the "spin quantum number" is simply called "spin" leaving its meaning as the unitless "spin quantum number" to be inferred from context.

When combined with the spin-statistics theorem, the spin of electrons results in the Pauli exclusion principle, which in turn underlies the periodic table of chemical elements.

Wolfgang Pauli was the first to propose the concept of spin, but he did not name it. In 1925, Ralph Kronig, George Uhlenbeck and Samuel Goudsmit at Leiden University suggested an erroneous physical interpretation of particles spinning around their own axis. The mathematical theory was worked out in depth by Pauli in 1927. When Paul Dirac derived his relativistic quantum mechanics in 1928, electron spin was an essential part of it.

Quantum Number

As the name suggests, spin was originally conceived as the rotation of a particle around some axis. This picture is correct so far as spin obeys the same mathematical laws as quantized angular momenta do. On the other hand, spin has some peculiar properties that distinguish it from orbital angular momenta:

- Spin quantum numbers may take half-integer values.
- Although the direction of its spin can be changed, an elementary particle cannot be made to spin faster or slower.
- The spin of a charged particle is associated with a magnetic dipole moment with a g-factor differing from 1. This could only occur classically if the internal charge of the particle were distributed differently from its mass.

The conventional definition of the spin quantum number, s, is $s = \frac{n}{2}$, where n can be any non-negative integer. Hence the allowed values of s are 0, $\frac{1}{2}$, 1, $\frac{3}{2}$, 2, etc. The value of s for an elementary particle depends only on the type of particle, and cannot be altered in any known way (in contrast to the spin direction described below). The spin angular momentum, S, of any physical system is quantized. The allowed values of S are

$$S = \hbar\sqrt{s(s+1)} = \frac{h}{4\pi}\sqrt{n(n+2)},$$

where h is the Planck constant and $\hbar = \frac{h}{2\pi}$ is the reduced Planck constant. In contrast, orbital angular momentum can only take on integer values of s; i.e., even-numbered values of n.

Fermions and Bosons

Those particles with half-integer spins, such as $\frac{1}{2}$, $\frac{3}{2}$, $\frac{5}{2}$, are known as fermions, while those particles with integer spins, such as 0, 1, 2, are known as bosons. The two families of particles obey different rules and broadly have different roles in the world around us. A key distinction between the two families is that fermions obey the Pauli exclusion principle; that is, there cannot be two identical fermions simultaneously having the same quantum numbers (meaning, roughly, having the same position, velocity and spin direction). In contrast, bosons obey the rules of Bose–Einstein statistics and have no such restriction, so they may "bunch together" even if in identical states. Also, composite particles can have spins different from their component particles. For example, a helium atom in the ground state has spin 0 and behaves like a boson, even though the quarks and electrons which make it up are all fermions.

This has profound consequences:

- Quarks and leptons (including electrons and neutrinos), which make up what is classically known as matter, are all fermions with spin $\frac{1}{2}$. The common idea that "matter takes up space" actually comes from the Pauli exclusion principle acting on these particles to prevent the fermions that make up matter from being in the same quantum state. Further compaction would require electrons to occupy the same energy states, and therefore a kind of pressure (sometimes known as degeneracy pressure of electrons) acts to resist the fermions being overly close. It is also this pressure which prevents stars collapsing inwardly, and which, when it finally gives way under immense gravitational pressure in a dying massive star, triggers inward collapse into a black hole.

  Elementary fermions with other spins ($\frac{3}{2}$, $\frac{5}{2}$, etc.) are not known to exist, as of 2014.

- Elementary particles which are thought of as carrying forces are all bosons with spin 1. They include the photon which carries the electromagnetic force, the gluon (strong force), and the W and Z bosons (weak force). The ability of bosons to occupy the same quantum state is used in the laser, which aligns many photons having the same quantum number (the same direction and frequency), superfluid liquid helium resulting from helium-4 atoms being bosons, and superconductivity where pairs of electrons (which individually are fermions) act as single composite bosons.

  Elementary bosons with other spins (0, 2, 3 etc.) were not historically known to exist, although they have received considerable theoretical treatment and are well established within their respective mainstream theories. In particular, theoreticians have proposed the graviton (predicted to exist by some quantum gravity theories) with spin 2, and the Higgs boson (explaining electroweak symmetry breaking) with spin 0. Since 2013, the Higgs boson with spin 0 has been considered proven to exist. It is the first scalar particle (spin 0) known to exist in nature.

Theoretical and experimental studies have shown that the spin possessed by elementary particles cannot be explained by postulating that they are made up of even smaller particles rotating about a common center of mass analogous to a classical electron radius; as far as can be determined at present, these elementary particles have no inner structure. The spin of an elementary particle is therefore seen as a truly intrinsic physical property, akin to the particle's electric charge and rest mass.

Spin-statistics Theorem

The fact that particles with half-integer spin (fermions) obey Fermi–Dirac statistics and the Pauli Exclusion Principle, and particles with integer spin (bosons) obey Bose–Einstein statistics, occupy "symmetric states", and thus can share quantum states, is known as the spin-statistics theorem. The theorem relies on both quantum mechanics and the theory of special relativity, and this connection between spin and statistics has been called "one of the most important applications of the special relativity theory".

Magnetic Moments

Particles with spin can possess a magnetic dipole moment, just like a rotating electrically charged body in classical electrodynamics. These magnetic moments can be experimentally observed in several ways, e.g. by the deflection of particles by inhomogeneous magnetic fields in a Stern–Gerlach experiment, or by measuring the magnetic fields generated by the particles themselves.

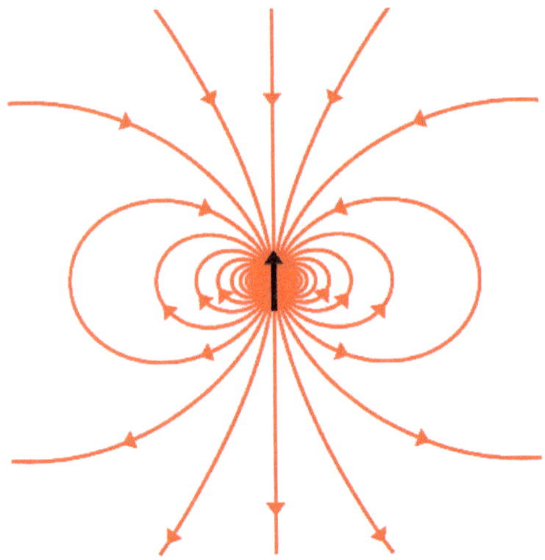

Schematic diagram depicting the spin of the neutron as the black arrow and magnetic field lines associated with the neutron magnetic moment. The neutron has a negative magnetic moment. While the spin of the neutron is upward in this diagram, the magnetic field lines at the center of the dipole are downward.

The intrinsic magnetic moment μ of a spin $\frac{1}{2}$ particle with charge q, mass m, and spin angular momentum S, is

$$\text{ì} = \frac{g_s q}{2m}\mathbf{S}$$

where the dimensionless quantity g_s is called the spin g-factor. For exclusively orbital rotations it would be 1 (assuming that the mass and the charge occupy spheres of equal radius).

The electron, being a charged elementary particle, possesses a nonzero magnetic moment. One of the triumphs of the theory of quantum electrodynamics is its accurate prediction of the electron g-factor, which has been experimentally determined to have the value −2.0023193043622(15), with the digits in parentheses denoting measurement uncertainty in the last two digits at one standard deviation. The value of 2 arises from the Dirac equation, a fundamental equation connecting the electron's spin with its electromagnetic properties, and the correction of 0.002319304... arises from the electron's interaction with the surrounding electromagnetic field, including its own field. Composite particles also possess magnetic moments associated with their spin. In particular, the neutron possesses a non-zero magnetic moment despite being electrically neutral. This fact was an early indication that the neutron is not an elementary particle. In fact, it is made up of quarks, which are electrically charged particles. The magnetic moment of the neutron comes from the spins of the individual quarks and their orbital motions.

Neutrinos are both elementary and electrically neutral. The minimally extended Standard Model that takes into account non-zero neutrino masses predicts neutrino magnetic moments of:

$$\mu_v \approx 3\times10^{-19}\,\mu_B \frac{m_v}{eV}$$

where the μ_v are the neutrino magnetic moments, m_v are the neutrino masses, and μ_B is the Bohr magneton. New physics above the electroweak scale could, however, lead to significantly higher neutrino magnetic moments. It can be shown in a model independent way that neutrino magnetic moments larger than about $10^{-14}\,\mu_B$ are unnatural, because they would also lead to large radiative contributions to the neutrino mass. Since the neutrino masses cannot exceed about 1 eV, these radiative corrections must then be assumed to be fine tuned to cancel out to a large degree.

The measurement of neutrino magnetic moments is an active area of research. As of 2001, the latest experimental results have put the neutrino magnetic moment at less than 1.2×10^{-10} times the electron's magnetic moment.

In ordinary materials, the magnetic dipole moments of individual atoms produce magnetic fields that cancel one another, because each dipole points in a random direction. Ferromagnetic materials below their Curie temperature, however, exhibit magnetic domains in which the atomic dipole moments are locally aligned, producing a macroscopic, non-zero magnetic field from the domain. These are the ordinary "magnets" with which we are all familiar.

In paramagnetic materials, the magnetic dipole moments of individual atoms spontaneously align with an externally applied magnetic field. In diamagnetic materials, on the other hand, the magnetic dipole moments of individual atoms spontaneously align oppositely to any externally applied magnetic field, even if it requires energy to do so.

The study of the behavior of such "spin models" is a thriving area of research in condensed matter physics. For instance, the Ising model describes spins (dipoles) that have only two possible states, up and down, whereas in the Heisenberg model the spin vector is allowed to point in any direction. These models have many interesting properties, which have led to interesting results in the theory of phase transitions.

Direction

Spin Projection Quantum Number and Multiplicity

In classical mechanics, the angular momentum of a particle possesses not only a magnitude (how fast the body is rotating), but also a direction (either up or down on the axis of rotation of the particle). Quantum mechanical spin also contains information about direction, but in a more subtle form. Quantum mechanics states that the component of angular momentum measured along any direction can only take on the values

$$S_i = \hbar s_i, \quad s_i \in \{-s, -(s-1), \ldots, s-1, s\}$$

where S_i is the spin component along the i-axis (either x, y, or z), s_i is the spin projection quantum number along the i-axis, and s is the principal spin quantum number. Conventionally the direction chosen is the z-axis:

$$S_z = \hbar s_z, \quad s_z \in \{-s, -(s-1), \ldots, s-1, s\}$$

where S_z is the spin component along the z-axis, s_z is the spin projection quantum number along the z-axis.

One can see that there are $2s + 1$ possible values of s_z. The number "$2s + 1$" is the multiplicity of the spin system. For example, there are only two possible values for a spin-$\frac{1}{2}$ particle: $s_z = +\frac{1}{2}$ and $s_z = -\frac{1}{2}$. These correspond to quantum states in which the spin is pointing in the +z or −z directions respectively, and are often referred to as "spin up" and "spin down". For a spin-3/2 particle, like a delta baryon, the possible values are $+\frac{3}{2}, +\frac{1}{2}, -\frac{1}{2}, -\frac{3}{2}$.

Vector

A single point in space can spin continuously without becoming tangled. Notice that after a 360-degree rotation, the spiral flips between clockwise and counterclockwise orientations. It returns to its original configuration after spinning a full 720 degrees.

For a given quantum state, one could think of a spin vector $\langle S \rangle$ whose components are the expectation values of the spin components along each axis, i.e., $\langle S \rangle = [\langle S_x \rangle, \langle S_y \rangle, \langle S_z \rangle]$. This vector then would describe the "direction" in which the spin is pointing, corresponding to the classical concept of the axis of rotation. It turns out that the spin vector is not very useful in actual quantum mechanical calculations, because it cannot be measured directly: sx, sy and sz cannot possess simultaneous definite values, because of a quantum uncertainty relation between them. However, for statistically large collections of particles that have been placed in the same pure quantum state, such as through the use of a Stern–Gerlach apparatus, the spin vector does have a well-defined experimental meaning: It specifies the direction in ordinary space in which a subsequent detector must be oriented in order to achieve the maximum possible probability (100%) of detecting every particle in the collection. For spin-1/2 particles, this maximum probability drops off smoothly as the angle between the spin vector and the detector increases, until at an angle of 180 degrees—that is, for detectors oriented in the opposite direction to the spin vector—the expectation of detecting particles from the collection reaches a minimum of 0%.

As a qualitative concept, the spin vector is often handy because it is easy to picture classically. For instance, quantum mechanical spin can exhibit phenomena analogous to classical gyroscopic effects. For example, one can exert a kind of "torque" on an electron by putting it in a magnetic field (the field acts upon the electron's intrinsic magnetic dipole moment). The result is that the spin vector undergoes precession, just like a classical gyroscope. This phenomenon is known as electron spin resonance (ESR). The equivalent behaviour of protons in atomic nuclei is used in nuclear magnetic resonance (NMR) spectroscopy and imaging.

Mathematically, quantum-mechanical spin states are described by vector-like objects known as spinors. There are subtle differences between the behavior of spinors and vectors under coordinate rotations. For example, rotating a spin-1/2 particle by 360 degrees does not bring it back to the same quantum state, but to the state with the opposite quantum phase; this is detectable, in principle, with interference experiments. To return the particle to its exact original state, one needs a 720-degree rotation. (The Plate trick and Mobius strip give non-quantum analogies.) A spin-zero particle can only have a single quantum state, even after torque is applied. Rotating a spin-2 particle 180 degrees can bring it back to the same quantum state and a spin-4 particle should be rotated 90 degrees to bring it back to the same quantum state. The spin-2 particle can be analogous to a straight stick that looks the same even after it is rotated 180 degrees and a spin 0 particle can be imagined as sphere, which looks the same after whatever angle it is turned through.

Mathematical Formulation

Operator

Spin obeys commutation relations analogous to those of the orbital angular momentum:

$$[S_j, S_k] = i\hbar\varepsilon_{jkl}S_l$$

where ε_{jkl} is the Levi-Civita symbol, and the expression is summed over the lth index. It follows (as with angular momentum) that the eigenvectors of S^2 and S_z (expressed as kets in the total S basis) are:

$$S^2 \,|\, s, m_s\rangle = \hbar^2 s(s+1) \,|\, s, m_s\rangle$$
$$S_z \,|\, s, m_s\rangle = \hbar m_s \,|\, s, m_s\rangle.$$

The spin raising and lowering operators acting on these eigenvectors give:

$$S_\pm \,|\, s, m_s\rangle = \hbar\sqrt{s(s+1) - m_s(m_s \pm 1)} \,|\, s, m_s \pm 1\rangle$$

where $S_\pm = S_x \pm i\, S_y$.

But unlike orbital angular momentum the eigenvectors are not spherical harmonics. They are not functions of θ and φ. There is also no reason to exclude half-integer values of s and m_s.

In addition to their other properties, all quantum mechanical particles possess an intrinsic spin (though it may have the intrinsic spin 0, too). The spin is quantized in units of the reduced Planck constant, such that the state function of the particle is, say, not $\psi = \psi(\mathbf{r})$, but $\psi = \psi(\mathbf{r},\sigma)$ where σ is out of the following discrete set of values:

$$\sigma \in \{-s\hbar, -(s-1)\hbar, \cdots, +(s-1)\hbar, +s\hbar\}.$$

One distinguishes bosons (integer spin) and fermions (half-integer spin). The total angular momentum conserved in interaction processes is then the sum of the orbital angular momentum and the spin.

Pauli Matrices

The quantum mechanical operators associated with spin-1/2 observables are:

$$\hat{S} = \frac{\hbar}{2}\sigma$$

where in Cartesian components:

$$S_x = \frac{\hbar}{2}\sigma_x, \quad S_y = \frac{\hbar}{2}\sigma_y, \quad S_z = \frac{\hbar}{2}\sigma_z.$$

For the special case of spin-1/2 particles, σ_x, σ_y and σ_z are the three Pauli matrices, given by:

$$\sigma_x = \begin{pmatrix} 0 & 1 \\ 1 & 0 \end{pmatrix} \quad \sigma_y = \begin{pmatrix} 0 & -i \\ i & 0 \end{pmatrix} \quad \sigma_z = \begin{pmatrix} 1 & 0 \\ 0 & -1 \end{pmatrix}.$$

Pauli Exclusion Principle

For systems of N identical particles this is related to the Pauli exclusion principle, which states that by interchanges of any two of the N particles one must have

$$\psi(\cdots \mathbf{r}_i, \sigma_i \cdots \mathbf{r}_j, \sigma_j \cdots) = (-1)^{2s}\psi(\cdots \mathbf{r}_j, \sigma_j \cdots \mathbf{r}_i, \sigma_i \cdots).$$

Thus, for bosons the prefactor $(-1)^{2s}$ will reduce to +1, for fermions to −1. In quantum mechanics all particles are either bosons or fermions. In some speculative relativistic quantum field theories "supersymmetric" particles also exist, where linear combinations of bosonic and fermionic components appear. In two dimensions, the prefactor $(-1)^{2s}$ can be replaced by any complex number of magnitude 1 such as in the anyon.

The above permutation postulate for N-particle state functions has most-important consequences in daily life, e.g. the periodic table of the chemical elements.

Rotations

As described above, quantum mechanics states that components of angular momentum measured along any direction can only take a number of discrete values. The most convenient quantum mechanical description of particle's spin is therefore with a set of complex numbers corresponding to amplitudes of finding a given value of projection of its intrinsic angular momentum on a given axis. For instance, for a spin-1/2 particle, we would need two numbers $a_{\pm 1/2}$, giving amplitudes of finding it with projection of angular momentum equal to $\hbar/2$ and $-\hbar/2$, satisfying the requirement.

$$\left|a_{\frac{1}{2}}\right|^2 + \left|a_{-\frac{1}{2}}\right|^2 = 1.$$

For a generic particle with spin s, we would need 2s + 1 such parameters. Since these numbers depend on the choice of the axis, they transform into each other non-trivially when this axis is rotated. It's clear that the transformation law must be linear, so we can represent it by associating a matrix with each rotation, and the product of two transformation matrices corresponding to rotations A and B must be equal (up to phase) to the matrix representing rotation AB. Further, rotations preserve the quantum mechanical inner product, and so should our transformation matrices:

$$\sum_{m=-j}^{j} a_m^* b_m = \sum_{m=-j}^{j}\left(\sum_{n=-j}^{j} U_{nm} a_n\right)^*\left(\sum_{k=-j}^{j} U_{km} b_k\right)$$

$$\sum_{n=-j}^{j}\sum_{k=-j}^{j} U_{np}^* U_{kq} = \delta_{pq}.$$

Mathematically speaking, these matrices furnish a unitary projective representation of the rotation group SO(3). Each such representation corresponds to a representation of the covering group of SO(3), which is SU(2). There is one n-dimensional irreducible representation of SU(2) for each dimension, though this representation is n-dimensional real for odd n and n-dimensional complex for even n (hence of real dimension 2n). For a rotation by angle θ in the plane with normal vector $\hat{\theta}$, U can be written.

$$U = e^{-\frac{i}{\hbar}\theta\cdot \mathrm{S}}$$

where $\theta = \theta\hat{\theta}$, and S is the vector of spin operators.

Working in the coordinate system where $\hat{\theta} = \hat{z}$, we would like to show that Sx and Sy are rotated into each other by the angle θ. Starting with Sx. Using units where $\hbar = 1$:

$$\begin{aligned} S_x \to U^\dagger S_x U &= e^{i\theta S_z} S_x e^{-i\theta S_z} \\ &= S_x + (i\theta)[S_z, S_x] + \left(\frac{1}{2!}\right)(i\theta)^2[S_z,[S_z,S_x]] + \left(\frac{1}{3!}\right)(i\theta)^3[S_z,[S_z,[S_z,S_x]]] + \cdots \end{aligned}$$

Using the spin operator commutation relations, we see that the commutators evaluate to $i\,S_y$ for the odd terms in the series, and to S_x for all of the even terms. Thus:

$$\begin{aligned} U^\dagger S_x U &= S_x\left[1 - \frac{\theta^2}{2!} + \ldots\right] - S_y\left[\theta - \frac{\theta^3}{3!}\cdots\right] \\ &= S_x \cos\theta - S_y \sin\theta \end{aligned}$$

as expected. Note that since we only relied on the spin operator commutation relations, this proof holds for any dimension (i.e., for any principal spin quantum number s).

A generic rotation in 3-dimensional space can be built by compounding operators of this type using Euler angles:

$$\mathcal{R}(\alpha,\beta,\gamma)=e^{-i\alpha S_z}e^{-i\beta S_y}e^{-i\gamma S_z}$$

An irreducible representation of this group of operators is furnished by the Wigner D-matrix:

$$D^s_{m'm}(\alpha,\beta,\gamma)\equiv\langle sm'|\mathcal{R}(\alpha,\beta,\gamma)|sm\rangle=e^{-im'\alpha}d^s_{m'm}(\beta)e^{-im\gamma},$$

where

$$d^s_{m'm}(\beta)=\langle sm'|e^{-i\beta s_y}|sm\rangle$$

is Wigner's small d-matrix. Note that for $\gamma = 2\pi$ and $\alpha = \beta = 0$; i.e., a full rotation about the z-axis, the Wigner D-matrix elements become

$$D^s_{m'm}(0,0,2\pi)=d^s_{m'm}(0)e^{-im2\pi}=\delta_{m'm}(-1)^{2m}.$$

Recalling that a generic spin state can be written as a superposition of states with definite m, we see that if s is an integer, the values of m are all integers, and this matrix corresponds to the identity operator. However, if s is a half-integer, the values of m are also all half-integers, giving $(-1)^{2m} = -1$ for all m, and hence upon rotation by 2π the state picks up a minus sign. This fact is a crucial element of the proof of the spin-statistics theorem.

Lorentz Transformations

We could try the same approach to determine the behavior of spin under general Lorentz transformations, but we would immediately discover a major obstacle. Unlike SO(3), the group of Lorentz transformations SO(3,1) is non-compact and therefore does not have any faithful, unitary, finite-dimensional representations.

In case of spin-1/2 particles, it is possible to find a construction that includes both a finite-dimensional representation and a scalar product that is preserved by this representation. We associate a 4-component Dirac spinor ψ with each particle. These spinors transform under Lorentz transformations according to the law.

$$\psi'=\exp\left(\tfrac{1}{8}\omega_{\mu\nu}[\gamma_\mu,\gamma_\nu]\right)\psi$$

where γ_ν are gamma matrices and $\omega_{\mu\nu}$ is an antisymmetric 4 × 4 matrix parametrizing the transformation. It can be shown that the scalar product.

$$\langle\psi|\phi\rangle=\bar{\psi}\phi=\psi^\dagger\gamma_0\phi$$

is preserved. It is not, however, positive definite, so the representation is not unitary.

Measurement of Spin Along the x-, y-, or z-axes

Each of the (Hermitian) Pauli matrices has two eigenvalues, +1 and −1. The corresponding normalized eigenvectors are:

$$\psi_{x+} = \frac{1}{\sqrt{2}}\begin{pmatrix}1\\1\end{pmatrix}, \quad \psi_{x-} = \frac{1}{\sqrt{2}}\begin{pmatrix}-1\\1\end{pmatrix},$$
$$\psi_{y+} = \frac{1}{\sqrt{2}}\begin{pmatrix}1\\i\end{pmatrix}, \quad \psi_{y-} = \frac{1}{\sqrt{2}}\begin{pmatrix}1\\-i\end{pmatrix},$$
$$\psi_{z+} = \begin{pmatrix}1\\0\end{pmatrix}, \quad \psi_{z-} = \begin{pmatrix}0\\1\end{pmatrix}.$$

By the postulates of quantum mechanics, an experiment designed to measure the electron spin on the x-, y-, or z-axis can only yield an eigenvalue of the corresponding spin operator (S_x, S_y or S_z) on that axis, i.e. $\hbar/2$ or $-\hbar/2$. The quantum state of a particle (with respect to spin), can be represented by a two component spinor:

$$\begin{pmatrix}a & bi\\c & di\end{pmatrix}$$

When the spin of this particle is measured with respect to a given axis (in this example, the x-axis), the probability that its spin will be measured as ħ/2 is just $|\langle\psi_{x+}|\psi\rangle|^2$. Correspondingly, the probability that its spin will be measured as −ħ/2 is just $|\langle\psi_{x-}|\psi\rangle|^2$. Following the measurement, the spin state of the particle will collapse into the corresponding eigenstate. As a result, if the particle's spin along a given axis has been measured to have a given eigenvalue, all measurements will yield the same eigenvalue (since $|\langle\psi_{x+}|\psi_{x+}\rangle|^2 = 1$, etc), provided that no measurements of the spin are made along other axes.

Measurement of Spin Along an Arbitrary Axis

The operator to measure spin along an arbitrary axis direction is easily obtained from the Pauli spin matrices. Let $u = (u_x, u_y, u_z)$ be an arbitrary unit vector. Then the operator for spin in this direction is simply

$$S_u = \frac{\hbar}{2}(u_x\sigma_x + u_y\sigma_y + u_z\sigma_z).$$

The operator S_u has eigenvalues of $\pm\frac{\hbar}{2}$, just like the usual spin matrices. This method of finding the operator for spin in an arbitrary direction generalizes to higher spin states, one takes the dot product of the direction with a vector of the three operators for the three x-, y-, z-axis directions.

A normalized spinor for spin-1/2 in the (u_x, u_y, u_z) direction (which works for all spin states except spin down where it will give 0/0), is:

$$\frac{1}{\sqrt{2+2u_z}}\begin{pmatrix}1+u_z\\u_x+iu_y\end{pmatrix}.$$

The above spinor is obtained in the usual way by diagonalizing the σ_u matrix and finding the eigenstates corresponding to the eigenvalues. In quantum mechanics, vectors are termed "normalized" when multiplied by a normalizing factor, which results in the vector having a length of unity.

Compatibility of Spin Measurements

Since the Pauli matrices do not commute, measurements of spin along the different axes are incompatible. This means that if, for example, we know the spin along the *x*-axis, and we then measure the spin along the *y*-axis, we have invalidated our previous knowledge of the *x*-axis spin. This can be seen from the property of the eigenvectors (i.e. eigenstates) of the Pauli matrices that:

$$\left|\langle\psi_{x\pm}|\psi_{y\pm}\rangle\right|^2 = \left|\langle\psi_{x\pm}|\psi_{z\pm}\rangle\right|^2 = \left|\langle\psi_{y\pm}|\psi_{z\pm}\rangle\right|^2 = \tfrac{1}{2}.$$

So when physicists measure the spin of a particle along the x-axis as, for example, ħ/2, the particle's spin state collapses into the eigenstate $|\psi_{x+}\rangle$. When we then subsequently measure the particle's spin along the y-axis, the spin state will now collapse into either $|\psi_{y+}\rangle$ or $|\psi_{y-}\rangle$, each with probability 1/2. Let us say, in our example, that we measure $-\frac{\hbar}{2}$. When we now return to measure the particle's spin along the x-axis again, the probabilities that we will measure $\frac{\hbar}{2}$ or $-\frac{\hbar}{2}$ are each 1/2 (i.e. they are $\left|\langle\psi_{x+}|\psi_{y-}\rangle\right|^2$ and $\left|\langle\psi_{x-}|\psi_{y-}\rangle\right|$ respectively). This implies that the original measurement of the spin along the x-axis is no longer valid, since the spin along the x-axis will now be measured to have either eigenvalue with equal probability.

Higher Spins

The spin-1/2 operator S = ħ/2σ form the fundamental representation of SU(2). By taking Kronecker products of this representation with itself repeatedly, one may construct all higher irreducible representations. That is, the resulting spin operators for higher spin systems in three spatial dimensions, for arbitrarily large s, can be calculated using this spin operator and ladder operators.

The resulting spin matrices for spin 1 are:

$$S_x = \frac{\hbar}{\sqrt{2}}\begin{pmatrix} 0 & 1 & 0 \\ 1 & 0 & 1 \\ 0 & 1 & 0 \end{pmatrix}$$

$$S_y = \frac{\hbar}{\sqrt{2}}\begin{pmatrix} 0 & -i & 0 \\ i & 0 & -i \\ 0 & i & 0 \end{pmatrix}$$

$$S_z = \hbar\begin{pmatrix} 1 & 0 & 0 \\ 0 & 0 & 0 \\ 0 & 0 & -1 \end{pmatrix}$$

for spin $\frac{3}{2}$ they are

$$S_x = \frac{\hbar}{2}\begin{pmatrix} 0 & \sqrt{3} & 0 & 0 \\ \sqrt{3} & 0 & 2 & 0 \\ 0 & 2 & 0 & \sqrt{3} \\ 0 & 0 & \sqrt{3} & 0 \end{pmatrix}$$

$$S_y = \frac{\hbar}{2}\begin{pmatrix} 0 & -i\sqrt{3} & 0 & 0 \\ i\sqrt{3} & 0 & -2i & 0 \\ 0 & 2i & 0 & -i\sqrt{3} \\ 0 & 0 & i\sqrt{3} & 0 \end{pmatrix}$$

$$S_z = \frac{\hbar}{2}\begin{pmatrix} 3 & 0 & 0 & 0 \\ 0 & 1 & 0 & 0 \\ 0 & 0 & -1 & 0 \\ 0 & 0 & 0 & -3 \end{pmatrix}$$

and for spin $\frac{5}{2}$ they are

$$S_x = \frac{\hbar}{2}\begin{pmatrix} 0 & \sqrt{5} & 0 & 0 & 0 & 0 \\ \sqrt{5} & 0 & 2\sqrt{2} & 0 & 0 & 0 \\ 0 & 2\sqrt{2} & 0 & 3 & 0 & 0 \\ 0 & 0 & 3 & 0 & 2\sqrt{2} & 0 \\ 0 & 0 & 0 & 2\sqrt{2} & 0 & \sqrt{5} \\ 0 & 0 & 0 & 0 & \sqrt{5} & 0 \end{pmatrix}$$

$$S_y = \frac{\hbar}{2}\begin{pmatrix} 0 & -i\sqrt{5} & 0 & 0 & 0 & 0 \\ i\sqrt{5} & 0 & -2i\sqrt{2} & 0 & 0 & 0 \\ 0 & 2i\sqrt{2} & 0 & -3i & 0 & 0 \\ 0 & 0 & 3i & 0 & -2i\sqrt{2} & 0 \\ 0 & 0 & 0 & 2i\sqrt{2} & 0 & -i\sqrt{5} \\ 0 & 0 & 0 & 0 & i\sqrt{5} & 0 \end{pmatrix}$$

$$S_z = \frac{\hbar}{2}\begin{pmatrix} 5 & 0 & 0 & 0 & 0 & 0 \\ 0 & 3 & 0 & 0 & 0 & 0 \\ 0 & 0 & 1 & 0 & 0 & 0 \\ 0 & 0 & 0 & -1 & 0 & 0 \\ 0 & 0 & 0 & 0 & -3 & 0 \\ 0 & 0 & 0 & 0 & 0 & -5 \end{pmatrix}.$$

The generalization of these matrices for arbitrary spin *s* is

$$\left(S_x\right)_{ab} = \frac{\hbar}{2}(\delta_{a,b+1} + \delta_{a+1,b})\sqrt{(s+1)(a+b-1)-ab}$$

$$\left(S_y\right)_{ab} = \frac{i\hbar}{2}(\delta_{a,b+1} - \delta_{a+1,b})\sqrt{(s+1)(a+b-1)-ab} \quad 1 \leq a,b \leq 2s+1$$

$$\left(S_z\right)_{ab} = \hbar(s+1-a)\delta_{a,b} = \hbar(s+1-b)\delta_{a,b}.$$

Also useful in the quantum mechanics of multiparticle systems, the general Pauli group G_n is defined to consist of all n-fold tensor products of Pauli matrices.

The analog formula of Euler's formula in terms of the Pauli matrices:

$$e^{i\theta(\hat{n}\cdot\sigma)} = I\cos\theta + i(\hat{n}\cdot\sigma)\sin\theta$$

for higher spins is tractable, but less simple.

Parity

In tables of the spin quantum number *s* for nuclei or particles, the spin is often followed by a "+" or "−". This refers to the parity with "+" for even parity (wave function unchanged by spatial inversion) and "−" for odd parity (wave function negated by spatial inversion). For example, the isotopes of bismuth.

Applications

Spin has important theoretical implications and practical applications. Well-established *direct* applications of spin include:

- Nuclear magnetic resonance (NMR) spectroscopy in chemistry;
- Electron spin resonance spectroscopy in chemistry and physics;
- Magnetic resonance imaging (MRI) in medicine, a type of applied NMR, which relies on proton spin density;
- Giant magnetoresistive (GMR) drive head technology in modern hard disks.

Electron spin plays an important role in magnetism, with applications for instance in computer memories. The manipulation of *nuclear spin* by radiofrequency waves (nuclear magnetic resonance) is important in chemical spectroscopy and medical imaging.

Spin-orbit coupling leads to the fine structure of atomic spectra, which is used in atomic clocks and in the modern definition of the second. Precise measurements of the g-factor of the electron have played an important role in the development and verification of quantum electrodynamics. *Photon spin* is associated with the polarization of light.

An emerging application of spin is as a binary information carrier in spin transistors. The original concept, proposed in 1990, is known as Datta-Das spin transistor. Electronics based on spin transistors are referred to as spintronics. The manipulation of spin in dilute magnetic semiconductor materials, such as metal-doped ZnO or TiO_2 imparts a further degree of freedom and has the potential to facilitate the fabrication of more efficient electronics.

There are many *indirect* applications and manifestations of spin and the associated Pauli exclusion principle, starting with the periodic table of chemistry.

History

Spin was first discovered in the context of the emission spectrum of alkali metals. In 1924 Wolfgang Pauli introduced what he called a "two-valued quantum degree of freedom" associated with the electron in the outermost shell. This allowed him to formulate the Pauli exclusion principle, stating that no two electrons can share the same quantum state at the same time.

Wolfgang Pauli lecturing

The physical interpretation of Pauli's "degree of freedom" was initially unknown. Ralph Kronig, one of Landé's assistants, suggested in early 1925 that it was produced by the self-rotation of the electron. When Pauli heard about the idea, he criticized it severely, noting that the electron's hypothetical surface would have to be moving faster than the speed of light in order for it to rotate quickly enough to produce the necessary angular momentum. This would violate the theory of relativity. Largely due to Pauli's criticism, Kronig decided not to publish his idea.

In the autumn of 1925, the same thought came to two Dutch physicists, George Uhlenbeck and Samuel Goudsmit at Leiden University. Under the advice of Paul Ehrenfest, they published their results. It met a favorable response, especially after Llewellyn Thomas managed to resolve a factor-of-two discrepancy between experimental results and Uhlenbeck and Goudsmit's calculations (and Kronig's unpublished results). This discrepancy was due to the orientation of the electron's tangent frame, in addition to its position.

Mathematically speaking, a fiber bundle description is needed. The tangent bundle effect is additive and relativistic; that is, it vanishes if c goes to infinity. It is one half of the value obtained without regard for the tangent space orientation, but with opposite sign. Thus the combined effect differs from the latter by a factor two (Thomas precession).

Despite his initial objections, Pauli formalized the theory of spin in 1927, using the modern theory of quantum mechanics invented by Schrödinger and Heisenberg. He pioneered the use of Pauli matrices as a representation of the spin operators, and introduced a two-component spinor wave-function.

Pauli's theory of spin was non-relativistic. However, in 1928, Paul Dirac published the Dirac equation, which described the relativistic electron. In the Dirac equation, a four-component spinor (known as a "Dirac spinor") was used for the electron wave-function. In 1940, Pauli proved the *spin-statistics theorem*, which states that fermions have half-integer spin and bosons have integer spin.

In retrospect, the first direct experimental evidence of the electron spin was the Stern–Gerlach experiment of 1922. However, the correct explanation of this experiment was only given in 1927.

Spontaneous Symmetry Breaking

Spontaneous symmetry breaking is a spontaneous process of symmetry breaking, by which a physical system in a symmetrical state ends up in an asymmetrical state. In particular, it can describe systems where the equations of motion or the Lagrangian obey symmetries, but the lowest-energy vacuum solutions do not exhibit that same symmetry.

Overview

In explicit symmetry breaking, if we consider two outcomes, the probability of a pair of outcomes can be different. By definition, spontaneous symmetry breaking requires the existence of a symmetrical probability distribution—any pair of outcomes has the same probability. In other words, the underlying laws are invariant under a symmetry transformation.

The system as a whole changes under such transformations.

Phases of matter, such as crystals, magnets, and conventional superconductors, as well as simple phase transitions can be described by spontaneous symmetry breaking. Notable exceptions include topological phases of matter like the fractional quantum Hall effect.

Examples

Mexican Hat Potential

Consider a symmetrical upward dome with a trough circling the bottom. If a ball is put at the very peak of the dome, the system is symmetrical with respect to a rotation around the center axis. But the ball may *spontaneously break* this symmetry by rolling down the dome into the trough, a point of lowest energy. Afterward, the ball has come to a rest at some fixed point on the perimeter. The dome and the ball retain their individual symmetry, but the system does not.

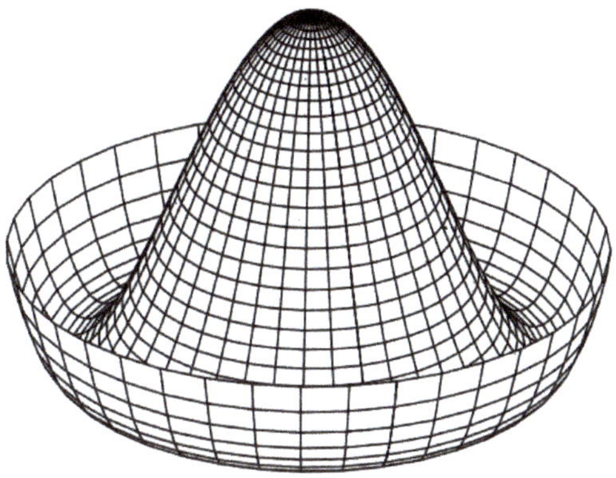

Graph of Goldstone's "Mexican hat" potential function V versus φ

In the simplest idealized relativistic model, the spontaneously broken symmetry is summarized through an illustrative scalar field theory. The relevant Lagrangian, which essentially dictates how a system behaves, can be split up into kinetic and potential terms,

$$\mathcal{L} = \partial^{\mu}\phi\partial_{\mu}\phi - V(\phi).$$

It is in this potential term $V(\Phi)$ that the symmetry breaking is triggered. An example of a potential, due to Jeffrey Goldstone is illustrated in the graph at the right.

$$V(\phi) = -10|\phi|^2 + |\phi|^4$$

This potential has an infinite number of possible minima (vacuum states) given by

$$\phi = \sqrt{5}e^{i\theta}.$$

for any real θ between 0 and 2π. The system also has an unstable vacuum state corresponding to $\Phi = 0$. This state has a U(1) symmetry. However, once the system falls into a specific stable vacuum state (amounting to a choice of θ), this symmetry will appear to be lost, or "spontaneously broken".

In fact, any other choice of θ would have exactly the same energy, implying the existence of a massless Nambu–Goldstone boson, the mode running around the circle at the minimum of this potential, and indicating there is some memory of the original symmetry in the Lagrangian.

Other Examples

- For ferromagnetic materials, the underlying laws are invariant under spatial rotations. Here, the order parameter is the magnetization, which measures the magnetic dipole density. Above the Curie temperature, the order parameter is zero, which is spatially invariant, and there is no symmetry breaking. Below the Curie temperature, however, the magnetization acquires a constant nonvanishing value, which points in a certain direction (in the idealized situation where we have full equilibrium; otherwise, translational symmetry gets broken as well). The residual rotational symmetries which leave the orientation of this vector invariant remain unbroken, unlike the other rotations which do not and are thus spontaneously broken.

- The laws describing a solid are invariant under the full Euclidean group, but the solid itself spontaneously breaks this group down to a space group. The displacement and the orientation are the order parameters.
- General relativity has a Lorentz symmetry, but in FRW cosmological models, the mean 4-velocity field defined by averaging over the velocities of the galaxies (the galaxies act like gas particles at cosmological scales) acts as an order parameter breaking this symmetry. Similar comments can be made about the cosmic microwave background.
- For the electroweak model, as explained earlier, a component of the Higgs field provides the order parameter breaking the electroweak gauge symmetry to the electromagnetic gauge symmetry. Like the ferromagnetic example, there is a phase transition at the electroweak temperature. The same comment about us not tending to notice broken symmetries suggests why it took so long for us to discover electroweak unification.
- In superconductors, there is a condensed-matter collective field ψ, which acts as the order parameter breaking the electromagnetic gauge symmetry.
- Take a thin cylindrical plastic rod and push both ends together. Before buckling, the system is symmetric under rotation, and so visibly cylindrically symmetric. But after buckling, it looks different, and asymmetric. Nevertheless, features of the cylindrical symmetry are still there: ignoring friction, it would take no force to freely spin the rod around, displacing the ground state in time, and amounting to an oscillation of vanishing frequency, unlike the radial oscillations in the direction of the buckle. This spinning mode is effectively the requisite Nambu–Goldstone boson.
- Consider a uniform layer of fluid over an infinite horizontal plane. This system has all the symmetries of the Euclidean plane. But now heat the bottom surface uniformly so that it becomes much hotter than the upper surface. When the temperature gradient becomes large enough, convection cells will form, breaking the Euclidean symmetry.
- Consider a bead on a circular hoop that is rotated about a vertical diameter. As the rotational velocity is increased gradually from rest, the bead will initially stay at its initial equilibrium point at the bottom of the hoop (intuitively stable, lowest gravitational potential). At a certain critical rotational velocity, this point will become unstable and the bead will jump to one of two other newly created equilibria, equidistant from the center. Initially, the system is symmetric with respect to the diameter, yet after passing the critical velocity, the bead ends up in one of the two new equilibrium points, thus breaking the symmetry.

Spontaneous Symmetry Breaking in Physics

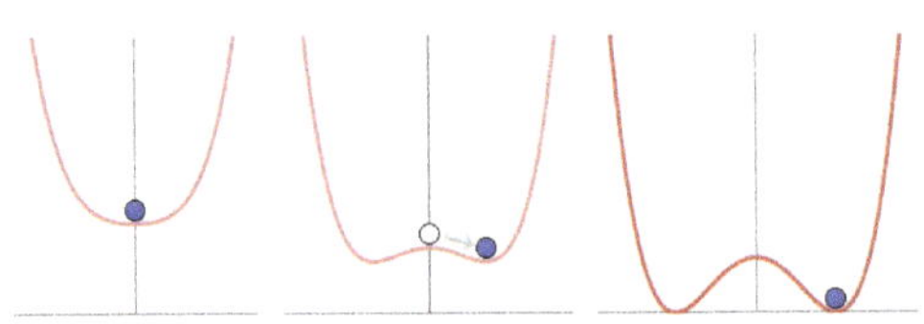

Spontaneous symmetry breaking illustrated: At high energy levels (*left*) the ball settles in the center, and the result is symmetrical. At lower energy levels (*right*), the overall "rules" remain symmetrical, but the symmetric "Mexican hat" enforces an asymmetric outcome, since eventually the ball must rest at some random spot on the bottom, "spontaneously", and not all others.

Particle Physics

In particle physics the force carrier particles are normally specified by field equations with gauge symmetry; their equations predict that certain measurements will be the same at any point in the field. For instance, field equations might predict that the mass of two quarks is constant. Solving the equations to find the mass of each quark might give two solutions. In one solution, quark A is heavier than quark B. In the second solution, quark B is heavier than quark A *by the same amount*. The symmetry of the equations is not reflected by the individual solutions, but it is reflected by the range of solutions.

An actual measurement reflects only one solution, representing a breakdown in the symmetry of the underlying theory. "Hidden" is a better term than "broken", because the symmetry is always there in these equations. This phenomenon is called *spontaneous* symmetry breaking (SSB) because *nothing* (that we know of) breaks the symmetry in the equations.

Chiral Symmetry

Chiral symmetry breaking is an example of spontaneous symmetry breaking affecting the chiral symmetry of the strong interactions in particle physics. It is a property of quantum chromodynamics, the quantum field theory describing these interactions, and is responsible for the bulk of the mass (over 99%) of the nucleons, and thus of all common matter, as it converts very light bound quarks into 100 times heavier constituents of baryons. The approximate Nambu–Goldstone bosons in this spontaneous symmetry breaking process are the pions, whose mass is an order of magnitude lighter than the mass of the nucleons. It served as the prototype and significant ingredient of the Higgs mechanism underlying the electroweak symmetry breaking.

Higgs Mechanism

The strong, weak, and electromagnetic forces can all be understood as arising from gauge symmetries. The Higgs mechanism, the spontaneous symmetry breaking of gauge symmetries, is an important component in understanding the superconductivity of metals and the origin of particle masses in the standard model of particle physics. One important consequence of the distinction between true symmetries and *gauge symmetries*, is that the spontaneous breaking of a gauge symmetry does not give rise to characteristic massless Nambu–Goldstone physical modes, but only massive modes, like the plasma mode in a superconductor, or the Higgs mode observed in particle physics.

In the standard model of particle physics, spontaneous symmetry breaking of the $SU(2) \times U(1)$ gauge symmetry associated with the electro-weak force generates masses for several particles, and separates the electromagnetic and weak forces. The W and Z bosons are the elementary particles that mediate the weak interaction, while the photon mediates the electromagnetic interaction. At energies much greater than 100 GeV all these particles behave in a similar manner. The Weinberg–Salam theory predicts that, at lower energies, this symmetry is broken so that the photon and the massive W and Z bosons emerge. In addition, fermions develop mass consistently.

Without spontaneous symmetry breaking, the Standard Model of elementary particle interactions requires the existence of a number of particles. However, some particles (the W and Z bosons) would then be predicted to be massless, when, in reality, they are observed to have mass. To overcome this, spontaneous symmetry breaking is augmented by the Higgs mechanism to give these particles mass. It also suggests the presence of a new particle, the Higgs boson, detected in 2012.

Superconductivity of metals is a condensed-matter analog of the Higgs phenomena, in which a condensate of Cooper pairs of electrons spontaneously breaks the U(1) gauge symmetry associated with light and electromagnetism.

Condensed Matter Physics

Most phases of matter can be understood through the lens of spontaneous symmetry breaking. For example, crystals are periodic arrays of atoms that are not invariant under all translations (only under a small subset of translations by a lattice vector). Magnets have north and south poles that are oriented in a specific direction, breaking rotational symmetry. In addition to these examples, there are a whole host of other symmetry-breaking phases of matter including nematic phases of liquid crystals, charge- and spin-density waves, superfluids and many others.

There are several known examples of matter that cannot be described by spontaneous symmetry breaking, including: topologically ordered phases of matter like fractional quantum Hall liquids, and spin-liquids. These states do not break any symmetry, but are distinct phases of matter. Unlike the case of spontaneous symmetry breaking, there is not a general framework for describing such states.

Continuous Symmetry

The ferromagnet is the canonical system which spontaneously breaks the continuous symmetry of the spins below the Curie temperature and at $h = 0$, where h is the external magnetic field. Below the Curie temperature the energy of the system is invariant under inversion of the magnetization $m(\mathbf{x})$ such that $m(\mathbf{x}) = -m(-\mathbf{x})$. The symmetry is spontaneously broken as $h \to 0$ when the Hamiltonian becomes invariant under the inversion transformation, but the expectation value is not invariant.

Spontaneously-symmetry-broken phases of matter are characterized by an order parameter that describes the quantity which breaks the symmetry under consideration. For example, in a magnet, the order parameter is the local magnetization.

Spontaneously breaking of a continuous symmetry is inevitably accompanied by gapless (meaning that these modes do not cost any energy to excite) Nambu–Goldstone modes associated with slow long-wavelength fluctuations of the order parameter. For example, vibrational modes in a crystal, known as phonons, are associated with slow density fluctuations of the crystal's atoms. The associated Goldstone mode for magnets are oscillating waves of spin known as spin-waves. For symmetry-breaking states, whose order parameter is not a conserved quantity, Nambu–Goldstone modes are typically massless and propagate at a constant velocity.

An important theorem, due to Mermin and Wagner, states that, at finite temperature, thermally activated fluctuations of Nambu–Goldstone modes destroy the long-range order, and prevent spontaneous symmetry breaking in one- and two-dimensional systems. Similarly, quantum fluctuations of the order parameter prevent most types of continuous symmetry breaking in one-dimensional systems even at zero temperature (an important exception is ferromagnets, whose order parameter, magnetization, is an exactly conserved quantity and does not have any quantum fluctuations).

Other long-range interacting systems such as cylindrical curved surfaces interacting via the Coulomb potential or Yukawa potential has been shown to break translational and rotational symmetries. It was shown, in the presence of a symmetric Hamiltonian, and in the limit of infinite volume, the system spontaneously adopts a chiral configuration, i.e. breaks mirror plane symmetry.

Dynamical Symmetry Breaking

Dynamical symmetry breaking (DSB) is a special form of spontaneous symmetry breaking where the ground state of the system has reduced symmetry properties compared to its theoretical description (Lagrangian).

Dynamical breaking of a global symmetry is a spontaneous symmetry breaking, that happens not at the (classical) tree level (i.e. at the level of the bare action), but due to quantum corrections (i.e. at the level of the effective action).

Dynamical breaking of a gauge symmetry is subtler. In the conventional spontaneous gauge symmetry breaking, there exists an unstable Higgs particle in the theory, which drives the vacuum to a symmetry-broken phase (e.g. Electroweak interaction). In dynamical gauge symmetry breaking, however, no unstable Higgs particle operates in the theory, but the bound states of the system itself provide the unstable fields that render the phase transition. For example, Bardeen, Hill, and Lindner published a paper which attempts to replace the conventional Higgs mechanism in the standard model, by a DSB that is driven by a bound state of top-antitop quarks (such models, where a composite particle plays the role of the Higgs boson, are often referred to as "Composite Higgs models"). Dynamical breaking of gauge symmetries is often due to creation of a fermionic condensate; for example the quark condensate, which is connected to the dynamical breaking of chiral symmetry in quantum chromodynamics. Conventional superconductivity is the paradigmatic example from the condensed matter side, where phonon-mediated attractions lead electrons to become bound in pairs and then condense, thereby breaking the electromagnetic gauge symmetry.

Generalisation and Technical Usage

For spontaneous symmetry breaking to occur, there must be a system in which there are several equally likely outcomes. The system as a whole is therefore symmetric with respect to these outcomes. However, if the system is sampled (i.e. if the system is actually used or interacted with in any way), a specific outcome must occur. Though the system as a whole is symmetric, it is never encountered with this symmetry, but only in one specific asymmetric state. Hence, the symmetry is said to be spontaneously broken in that theory. Nevertheless, the fact that each outcome is equally likely is a reflection of the underlying symmetry, which is thus often dubbed "hidden symmetry", and has crucial formal consequences.

When a theory is symmetric with respect to a symmetry group, but requires that one element of the group be distinct, then spontaneous symmetry breaking has occurred. The theory must not dictate *which* member is distinct, only that *one is*. From this point on, the theory can be treated as if this element actually is distinct, with the proviso that any results found in this way must be resymmetrized, by taking the average of each of the elements of the group being the distinct one.

The crucial concept in physics theories is the order parameter. If there is a field (often a background field) which acquires an expectation value (not necessarily a *vacuum* expectation value) which is not invariant under the symmetry in question, we say that the system is in the ordered phase, and the symmetry is spontaneously broken. This is because other subsystems interact with the order parameter, which specifies a "frame of reference" to be measured against. In that case, the vacuum state does not obey the initial symmetry (which would keep it invariant, in the linearly realized Wigner mode in which it would be a singlet), and, instead changes under the (hidden) symmetry, now implemented in the (nonlinear) Nambu–Goldstone mode. Normally, in the absence of the Higgs mechanism, massless Goldstone bosons arise.

The symmetry group can be discrete, such as the space group of a crystal, or continuous (e.g., a Lie group), such as the rotational symmetry of space. However, if the system contains only a single spatial dimension, then only discrete symmetries may be broken in a vacuum state of the full quantum theory, although a classical solution may break a continuous symmetry.

Nobel Prize

On October 7, 2008, the Royal Swedish Academy of Sciences awarded the 2008 Nobel Prize in Physics to three scientists for their work in subatomic physics symmetry breaking. Yoichiro Nambu, of the University of Chicago, won half of the prize for the discovery of the mechanism of spontaneous broken symmetry in the context of the strong interactions, specifically chiral symmetry breaking. Physicists Makoto Kobayashi and Toshihide Maskawa shared the other half of the prize for discovering the origin of the explicit breaking of CP symmetry in the weak interactions. This origin is ultimately reliant on the Higgs mechanism, but, so far understood as a "just so" feature of Higgs couplings, not a spontaneously broken symmetry phenomenon.

Magnetism

A magnetic quadrupole

Magnetism is a class of physical phenomena that are mediated by magnetic fields. Electric currents and the magnetic moments of elementary particles give rise to a magnetic field, which acts on other currents and magnetic moments. The most familiar effects occur in ferromagnetic materials, which are strongly attracted by magnetic fields and can be magnetized to become permanent magnets, producing magnetic fields themselves. Only a few substances are ferromagnetic; the most common ones are iron, nickel and cobalt and their alloys. The prefix *ferro-* refers to iron, because permanent magnetism was first observed in lodestone, a form of natural iron ore called magnetite, Fe_3O_4.

Although ferromagnetism is responsible for most of the effects of magnetism encountered in everyday life, all other materials are influenced to some extent by a magnetic field, by several other types of magnetism. Paramagnetic substances such as aluminum and oxygen are weakly attracted to an applied magnetic field; diamagnetic substances such as copper and carbon are weakly repelled; while antiferromagnetic materials such as chromium and spin glasses have a more complex relationship with a magnetic field. The force of a magnet on paramagnetic, diamagnetic, antiferromagnetic materials is usually too weak to be felt, and can be detected only by laboratory instruments, so in everyday life these substances are often described as non-magnetic.

The magnetic state (or magnetic phase) of a material depends on temperature and other variables such as pressure and the applied magnetic field. A material may exhibit more than one form of magnetism as these variables change.

History

Lodestone, a natural magnet, attracting iron nails. Ancient humans discovered the property of magnetism from lodestone.

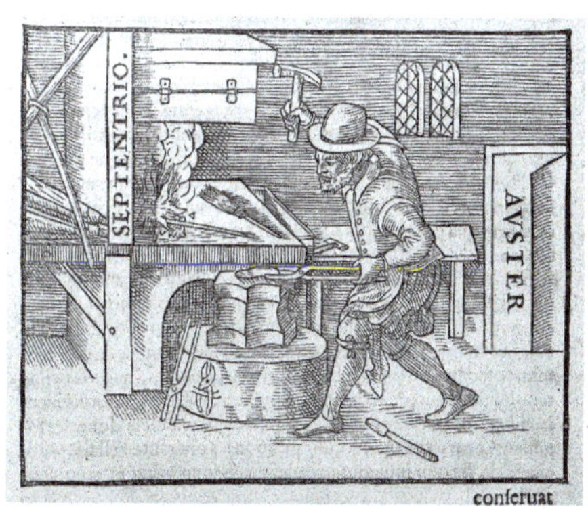

An illustration from Gilbert's 1600 *De Magnete* showing one of the earliest methods of making a magnet. A blacksmith holds a piece of red-hot iron in a north-south direction and hammers it as it cools. The magnetic field of the Earth aligns the domains, leaving the iron a weak magnet.

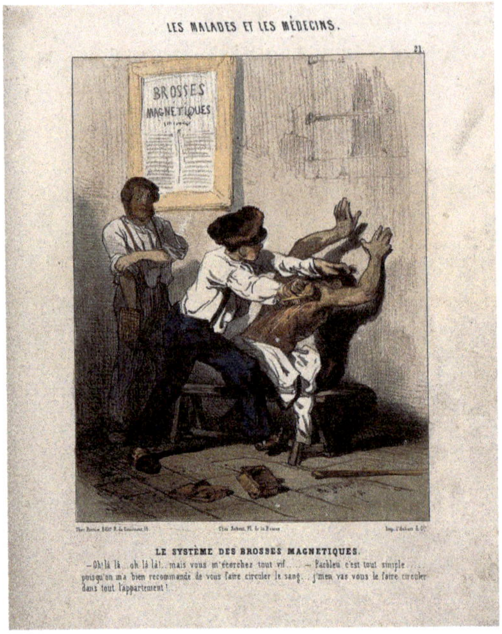

Drawing of a medical treatment using magnetic brushes. Charles Jacque 1843, France.

Magnetism was first discovered in the ancient world, when people noticed that lodestones, naturally magnetized pieces of the mineral magnetite, could attract iron. The word *magnet* comes from the Greek term *magnētis lithos*, "the Magnesian stone, lodestone." In ancient Greece, Aristotle attributed the first of what could be called a scientific discussion of magnetism to the philosopher Thales of Miletus, who lived from about 625 BC to about 545 BC. Around the same time, in ancient India, the Indian surgeon Sushruta was the first to make use of the magnet for surgical purposes.

In ancient China, the earliest literary reference to magnetism lies in a 4th-century BC book named after its author, *The Sage of Ghost Valley*. The 2nd-century BC annals, *Lüshi Chunqiu*, also notes: "The lodestone makes iron approach, or it attracts it." The earliest mention of the attraction of a needle is in a 1st-century work *Lunheng* (*Balanced Inquiries*): "A lodestone attracts a needle." The 11th-century Chinese scientist Shen Kuo was the first person to write – in the *Dream Pool Essays* – of the magnetic needle compass and that it improved the accuracy of navigation by employing the astronomical concept of true north. By the 12th century the Chinese were known to use the lodestone compass for navigation. They sculpted a directional spoon from lodestone in such a way that the handle of the spoon always pointed south.

Alexander Neckam, by 1187, was the first in Europe to describe the compass and its use for navigation. In 1269, Peter Peregrinus de Maricourt wrote the *Epistola de magnete*, the first extant treatise describing the properties of magnets. In 1282, the properties of magnets and the dry compass were discussed by Al-Ashraf, a Yemeni physicist, astronomer, and geographer.

In 1600, William Gilbert published his *De Magnete, Magneticisque Corporibus, et de Magno Magnete Tellure* (*On the Magnet and Magnetic Bodies, and on the Great Magnet the Earth*). In this work he describes many of his experiments with his model earth called the terrella. From his experiments, he concluded that the Earth was itself magnetic and that this was the reason compasses pointed north (previously, some believed that it was the pole star (Polaris) or a large magnetic island on the north pole that attracted the compass).

An understanding of the relationship between electricity and magnetism began in 1819 with work by Hans Christian Ørsted, a professor at the University of Copenhagen, who discovered by the accidental twitching of a compass needle near a wire that an electric current could create a magnetic field. This landmark experiment is known as Ørsted's Experiment. Several other experiments followed, with André-Marie Ampère, who in 1820 discovered that the magnetic field circulating in a closed-path was related to the current flowing through the perimeter of the path; Carl Friedrich Gauss; Jean-Baptiste Biot and Félix Savart, both of whom in 1820 came up with the Biot–Savart law giving an equation for the magnetic field from a current-carrying wire; Michael Faraday, who in 1831 found that a time-varying magnetic flux through a loop of wire induced a voltage, and others finding further links between magnetism and electricity. James Clerk Maxwell synthesized and expanded these insights into Maxwell's equations, unifying electricity, magnetism, and optics into the field of electromagnetism. In 1905, Einstein used these laws in motivating his theory of special relativity, requiring that the laws held true in all inertial reference frames.

Electromagnetism has continued to develop into the 21st century, being incorporated into the more fundamental theories of gauge theory, quantum electrodynamics, electroweak theory, and finally the standard model.

Sources of Magnetism

Magnetism, at its root, arises from two sources:

1. Electric current.
2. Spin magnetic moments of elementary particles. The magnetic moments of the nuclei of atoms are typically thousands of times smaller than the electrons' magnetic moments, so they are negligible in the context of the magnetization of materials. Nuclear magnetic moments are nevertheless very important in other contexts, particularly in nuclear magnetic resonance (NMR) and magnetic resonance imaging (MRI).

Ordinarily, the enormous number of electrons in a material are arranged such that their magnetic moments (both orbital and intrinsic) cancel out. This is due, to some extent, to electrons combining into pairs with opposite intrinsic magnetic moments as a result of the Pauli exclusion principle, or combining into filled subshells with zero net orbital motion. In both cases, the electron arrangement is so as to exactly cancel the magnetic moments from each electron. Moreover, even when the electron configuration *is* such that there are unpaired electrons and/or non-filled subshells, it is often the case that the various electrons in the solid will contribute magnetic moments that point in different, random directions, so that the material will not be magnetic.

Sometimes, either spontaneously, or owing to an applied external magnetic field—each of the electron magnetic moments will be, on average, lined up. A suitable material can then produce a strong net magnetic field.

The magnetic behavior of a material depends on its structure, particularly its electron configuration, for the reasons mentioned above, and also on the temperature. At high temperatures, random thermal motion makes it more difficult for the electrons to maintain alignment.

Materials

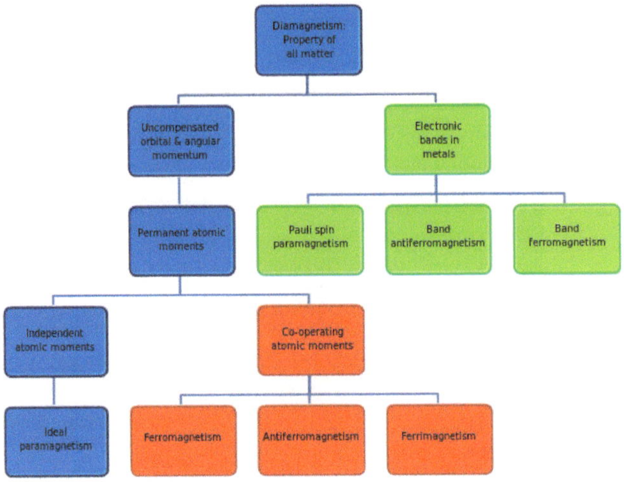

Hierarchy of types of magnetism.

Diamagnetism

Diamagnetism appears in all materials, and is the tendency of a material to oppose an applied magnetic field, and therefore, to be repelled by a magnetic field. However, in a material with paramagnetic properties (that is, with a tendency to enhance an external magnetic field), the paramagnetic behavior dominates. Thus, despite its universal occurrence, diamagnetic behavior is observed only in a purely diamagnetic material. In a diamagnetic material, there are no unpaired electrons, so the intrinsic electron magnetic moments cannot produce any bulk effect. In these cases, the magnetization arises from the electrons' orbital motions, which can be understood classically as follows:

When a material is put in a magnetic field, the electrons circling the nucleus will experience, in addition to their Coulomb attraction to the nucleus, a Lorentz force from the magnetic field. Depending on which direction the electron is orbiting, this force may increase the centripetal force on the electrons, pulling them in towards the nucleus, or it may decrease the force, pulling them away from the nucleus. This effect systematically increases the orbital magnetic moments that were aligned opposite the field, and decreases the ones aligned parallel to the field (in accordance with Lenz's law). This results in a small bulk magnetic moment, with an opposite direction to the applied field.

Note that this description is meant only as a heuristic; a proper understanding requires a quantum-mechanical description.

Note that all materials undergo this orbital response. However, in paramagnetic and ferromagnetic substances, the diamagnetic effect is overwhelmed by the much stronger effects caused by the unpaired electrons.

Paramagnetism

In a paramagnetic material there are *unpaired electrons*, i.e. atomic or molecular orbitals with exactly one electron in them. While paired electrons are required by the Pauli exclusion principle to have their intrinsic ('spin') magnetic moments pointing in opposite directions, causing their magnetic fields to cancel out, an unpaired electron is free to align its magnetic moment in any direction. When an external magnetic field is applied, these magnetic moments will tend to align themselves in the same direction as the applied field, thus reinforcing it.

Ferromagnetism

Tip of permanent magnet with coins demonstrating ferromagnetism

A ferromagnet, like a paramagnetic substance, has unpaired electrons. However, in *addition* to the electrons' intrinsic magnetic moment's tendency to be parallel to an *applied field*, there is also in these materials a tendency for these magnetic moments to orient parallel to *each other* to maintain a lowered-energy state. Thus, even in the absence of an applied field, the magnetic moments of the electrons in the material spontaneously line up parallel to one another.

Every ferromagnetic substance has its own individual temperature, called the Curie temperature, or Curie point, above which it loses its ferromagnetic properties. This is because the thermal tendency to disorder overwhelms the energy-lowering due to ferromagnetic order.

Ferromagnetism only occurs in a few substances; the common ones are iron, nickel, cobalt, their alloys, and some alloys of rare earth metals.

Magnetic Domains

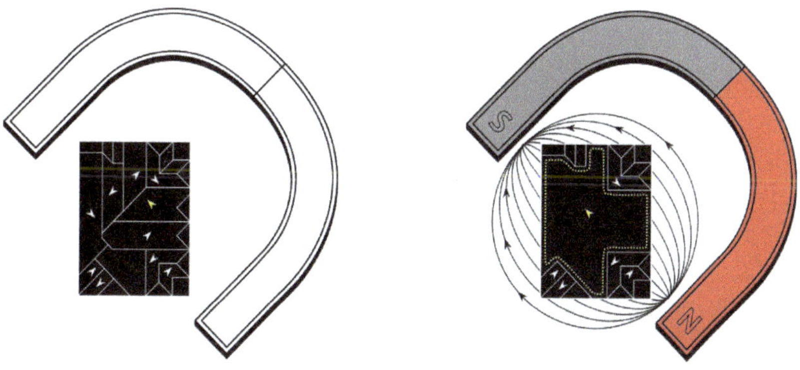

Left: Magnetic domains boundaries (white lines) in ferromagnetic material (black rectangle). Right: Effect of a magnet on the domains.

The magnetic moments of atoms in a ferromagnetic material cause them to behave something like tiny permanent magnets. They stick together and align themselves into small regions of more or less uniform alignment called magnetic domains or Weiss domains. Magnetic domains can be observed with a magnetic force microscope to reveal magnetic domain boundaries that resemble white lines in the sketch. There are many scientific experiments that can physically show magnetic fields.

When a domain contains too many molecules, it becomes unstable and divides into two domains aligned in opposite directions so that they stick together more stably as shown at the right.

When exposed to a magnetic field, the domain boundaries move so that the domains aligned with the magnetic field grow and dominate the structure (dotted yellow area) as shown at the left. When the magnetizing field is removed, the domains may not return to an unmagnetized state. This results in the ferromagnetic material's being magnetized, forming a permanent magnet.

When magnetized strongly enough that the prevailing domain overruns all others to result in only one single domain, the material is magnetically saturated. When a magnetized ferromagnetic material is heated to the Curie point temperature, the molecules are agitated to the point that the magnetic domains lose the organization and the magnetic properties they cause cease. When the material is cooled, this domain alignment structure spontaneously returns, in a manner roughly analogous to how a liquid can freeze into a crystalline solid.

Antiferromagnetism

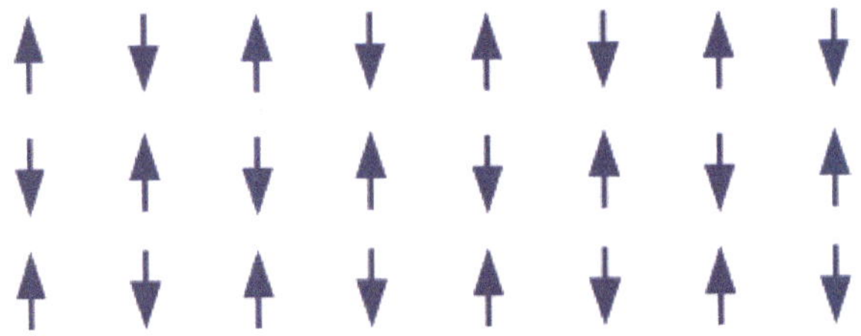

Antiferromagnetic ordering

In an antiferromagnet, unlike a ferromagnet, there is a tendency for the intrinsic magnetic moments of neighboring valence electrons to point in *opposite* directions. When all atoms are arranged in a substance so that each neighbor is 'anti-aligned', the substance is antiferromagnetic. Antiferromagnets have a zero net magnetic moment, meaning no field is produced by them. Antiferromagnets are less common compared to the other types of behaviors, and are mostly observed at low temperatures. In varying temperatures, antiferromagnets can be seen to exhibit diamagnetic and ferromagnetic properties.

In some materials, neighboring electrons want to point in opposite directions, but there is no geometrical arrangement in which *each* pair of neighbors is anti-aligned. This is called a spin glass, and is an example of geometrical frustration.

Ferrimagnetism

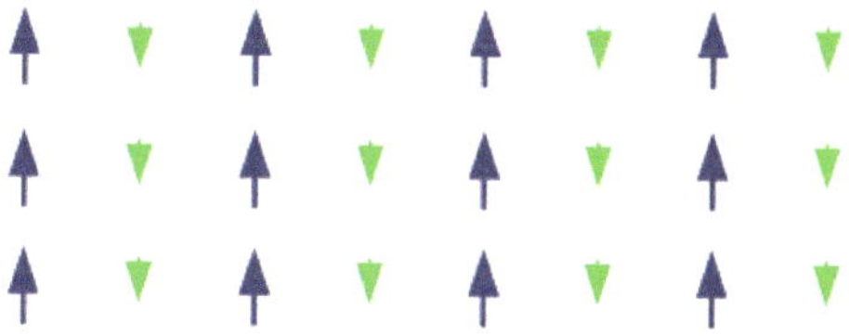

Ferrimagnetic ordering

Like ferromagnetism, ferrimagnets retain their magnetization in the absence of a field. However, like antiferromagnets, neighboring pairs of electron spins tend to point in opposite directions. These two properties are not contradictory, because in the optimal geometrical arrangement, there is more magnetic moment from the sublattice of electrons that point in one direction, than from the sublattice that points in the opposite direction.

Most ferrites are ferrimagnetic. The first discovered magnetic substance, magnetite, is a ferrite and was originally believed to be a ferromagnet; Louis Néel disproved this, however, after discovering ferrimagnetism.

Superparamagnetism

When a ferromagnet or ferrimagnet is sufficiently small, it acts like a single magnetic spin that is subject to Brownian motion. Its response to a magnetic field is qualitatively similar to the response of a paramagnet, but much larger.

Other Types of Magnetism

- Metamagnetism
- Molecule-based magnet
- Spin glass

Electromagnet

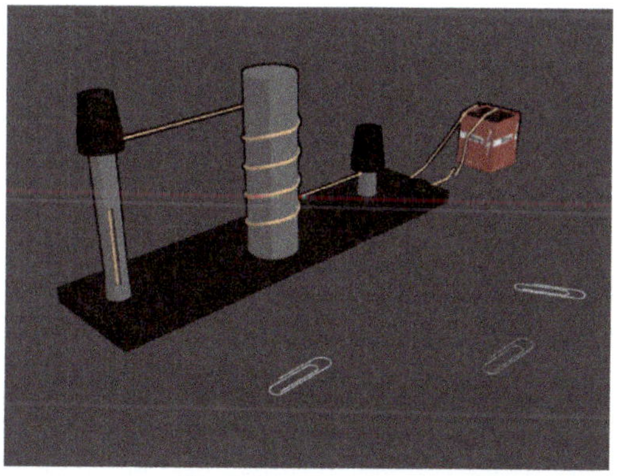

An electromagnet attracts paper clips when current is applied creating a magnetic field. The electromagnet loses them when current and magnetic field are removed.

An electromagnet is a type of magnet in which the magnetic field is produced by an electric current. The magnetic field disappears when the current is turned off. Electromagnets usually consist of a large number of closely spaced turns of wire that create the magnetic field. The wire turns are often wound around a magnetic core made from a ferromagnetic or ferrimagnetic material such as iron; the magnetic core concentrates the magnetic flux and makes a more powerful magnet.

The main advantage of an electromagnet over a permanent magnet is that the magnetic field can be quickly changed by controlling the amount of electric current in the winding. However, unlike a permanent magnet that needs no power, an electromagnet requires a continuous supply of current to maintain the magnetic field.

Electromagnets are widely used as components of other electrical devices, such as motors, generators, relays, loudspeakers, hard disks, MRI machines, scientific instruments, and magnetic separation equipment. Electromagnets are also employed in industry for picking up and moving heavy iron objects such as scrap iron and steel. Electromagnetism was discovered in 1820.

Magnetism, Electricity, and Special Relativity

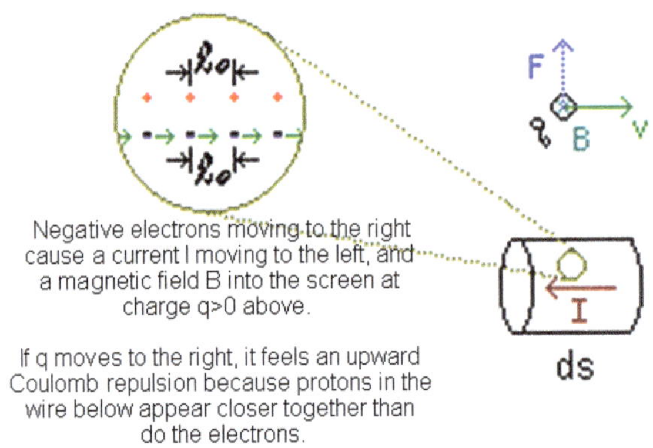

Magnetism from length-contraction

As a consequence of Einstein's theory of special relativity, electricity and magnetism are fundamentally interlinked. Both magnetism lacking electricity, and electricity without magnetism, are inconsistent with special relativity, due to such effects as length contraction, time dilation, and the fact that the magnetic force is velocity-dependent. However, when both electricity and magnetism are taken into account, the resulting theory (electromagnetism) is fully consistent with special relativity. In particular, a phenomenon that appears purely electric or purely magnetic to one observer may be a mix of both to another, or more generally the relative contributions of electricity and magnetism are dependent on the frame of reference. Thus, special relativity "mixes" electricity and magnetism into a single, inseparable phenomenon called electromagnetism, analogous to how relativity "mixes" space and time into spacetime.

All observations on electromagnetism apply to what might be considered to be primarily magnetism, e.g. perturbations in the magnetic field are necessarily accompanied by a nonzero electric field, and propagate at the speed of light.

Magnetic Fields in a Material

In a vacuum,

$$\mathbf{B} = \mu_0 \mathbf{H},$$

where μ_o is the vacuum permeability.

In a material,

$$\mathbf{B} = \mu_0 (\mathbf{H} + \mathbf{M}).$$

The quantity $\mu_o M$ is called *magnetic polarization.*

If the field H is small, the response of the magnetization M in a diamagnet or paramagnet is approximately linear:

$$\mathbf{M} = \chi \mathbf{H},$$

the constant of proportionality being called the magnetic susceptibility. If so,

$$\mu_0 (\mathbf{H} + \mathbf{M}) = \mu_0 (1 + \chi) \mathbf{H} = \mu_r \mu_0 \mathbf{H} = \mu \mathbf{H}.$$

In a hard magnet such as a ferromagnet, M is not proportional to the field and is generally nonzero even when H is zero.

Magnetic Force

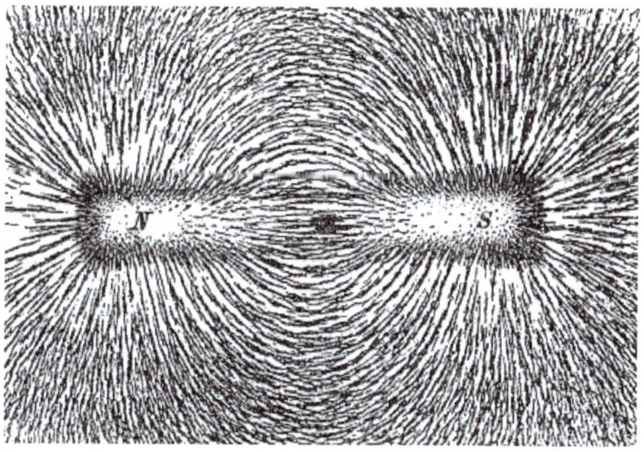

Magnetic lines of force of a bar magnet shown by iron filings on paper

The phenomenon of magnetism is "mediated" by the magnetic field. An electric current or magnetic dipole creates a magnetic field, and that field, in turn, imparts magnetic forces on other particles that are in the fields.

Maxwell's equations, which simplify to the Biot–Savart law in the case of steady currents, describe the origin and behavior of the fields that govern these forces. Therefore, magnetism is seen whenever electrically charged particles are in motion—for example, from movement of electrons in an electric current, or in certain cases from the orbital motion of electrons around an atom's nucleus. They also arise from "intrinsic" magnetic dipoles arising from quantum-mechanical spin.

The same situations that create magnetic fields—charge moving in a current or in an atom, and intrinsic magnetic dipoles—are also the situations in which a magnetic field has an effect, creating a force. Following is the formula for moving charge; for the forces on an intrinsic dipole.

When a charged particle moves through a magnetic field B, it feels a Lorentz force F given by the cross product:

$$\mathbf{F} = q(\mathbf{v} \times \mathbf{B})$$

where

q is the electric charge of the particle, and

v is the velocity vector of the particle

Because this is a cross product, the force is perpendicular to both the motion of the particle and the magnetic field. It follows that the magnetic force does no work on the particle; it may change the direction of the particle's movement, but it cannot cause it to speed up or slow down. The magnitude of the force is

$$F = qvB \sin \theta$$

where θ is the angle between v and B.

One tool for determining the direction of the velocity vector of a moving charge, the magnetic field, and the force exerted is labeling the index finger "V", the middle finger "B", and the thumb "F" with your right hand. When making a gun-like configuration, with the middle finger crossing under the index finger, the fingers represent the velocity vector, magnetic field vector, and force vector, respectively.

Magnetic Dipoles

A very common source of magnetic field found in nature is a dipole, with a "South pole" and a "North pole", terms dating back to the use of magnets as compasses, interacting with the Earth's magnetic field to indicate North and South on the globe. Since opposite ends of magnets are attracted, the north pole of a magnet is attracted to the south pole of another magnet. The Earth's North Magnetic Pole (currently in the Arctic Ocean, north of Canada) is physically a south pole, as it attracts the north pole of a compass. A magnetic field contains energy, and physical systems move toward configurations with lower energy. When diamagnetic material is placed in a magnetic field, a *magnetic dipole* tends to align itself in opposed polarity to that field, thereby lowering the net field strength. When ferromagnetic material is placed within a magnetic field, the magnetic dipoles align to the applied field, thus expanding the domain walls of the magnetic domains.

Magnetic Monopoles

Since a bar magnet gets its ferromagnetism from electrons distributed evenly throughout the bar, when a bar magnet is cut in half, each of the resulting pieces is a smaller bar magnet. Even though a magnet is said to have a north pole and a south pole, these two poles cannot be separated from each other. A monopole—if such a thing exists—would be a new and fundamentally different kind of magnetic object. It would act as an isolated north pole, not attached to a south pole, or vice versa. Monopoles would carry "magnetic charge" analogous to electric charge. Despite systematic searches since 1931, as of 2010, they have never been observed, and could very well not exist.

Nevertheless, some theoretical physics models predict the existence of these magnetic monopoles. Paul Dirac observed in 1931 that, because electricity and magnetism show a certain symmetry, just as quantum theory predicts that individual positive or negative electric charges can be observed without the opposing charge, isolated South or North magnetic poles should be observable. Using quantum theory Dirac showed that if magnetic monopoles exist, then one could explain the quantization of electric charge—that is, why the observed elementary particles carry charges that are multiples of the charge of the electron.

Certain grand unified theories predict the existence of monopoles which, unlike elementary particles, are solitons (localized energy packets). The initial results of using these models to estimate the number of monopoles created in the big bang contradicted cosmological observations—the monopoles would have been so plentiful and massive that they would have long since halted the expansion of the universe. However, the idea of inflation (for which this problem served as a partial motivation) was successful in solving this problem, creating models in which monopoles existed but were rare enough to be consistent with current observations.

Quantum-mechanical Origin of Magnetism

In principle all kinds of magnetism originate from specific quantum-mechanical phenomena (e.g. Mathematical formulation of quantum mechanics, in particular the chapters on spin and on the Pauli principle). A successful model was developed already in 1927, by Walter Heitler and Fritz London, who derived quantum-mechanically, how hydrogen molecules are formed from hydrogen atoms, i.e. from the atomic hydrogen orbitals u_A and u_B centered at the nuclei A and B. That this leads to magnetism is not at all obvious, but will be explained in the following.

According to the Heitler-London theory, so-called two-body molecular σ-orbitals are formed, namely the resulting orbital is:

$$\psi(\mathbf{r}_1,\mathbf{r}_2)=\frac{1}{\sqrt{2}}\left(u_A(\mathbf{r}_1)u_B(\mathbf{r}_2)+u_B(\mathbf{r}_1)u_A(\mathbf{r}_2)\right)$$

Here the last product means that a first electron, r_1, is in an atomic hydrogen-orbital centered at the second nucleus, whereas the second electron runs around the first nucleus. This "exchange" phenomenon is an expression for the quantum-mechanical property that particles with identical properties cannot be distinguished. It is specific not only for the formation of chemical bonds, but as we will see, also for magnetism, i.e. in this connection the term exchange interaction arises, a term which is essential for the origin of magnetism, and which is stronger, roughly by factors 100 and even by 1000, than the energies arising from the electrodynamic dipole-dipole interaction.

As for the spin function $\chi(s_1, s_2)$, which is responsible for the magnetism, we have the already mentioned Pauli's principle, namely that a symmetric orbital (i.e. with the + sign as above) must be multiplied with an antisymmetric spin function (i.e. with a – sign), and vice versa. Thus:

$$\chi(s_1, s_2) = \frac{1}{\sqrt{2}}\left(\alpha(s_1)\beta(s_2) - \beta(s_1)\alpha(s_2)\right),$$

I.e., not only u_A and u_B must be substituted by α and β, respectively (the first entity means "spin up", the second one "spin down"), but also the sign + by the – sign, and finally ri by the discrete values si (= ±½); thereby we have $\alpha(+1/2) = \beta(-1/2) = 1$ and $\alpha(-1/2) = \beta(+1/2) = 0$. The "singlet state", i.e. the – sign, means: the spins are antiparallel, i.e. for the solid we have antiferromagnetism, and for two-atomic molecules one has diamagnetism. The tendency to form a (homoeopolar) chemical bond (this means: the formation of a symmetric molecular orbital, i.e. with the + sign) results through the Pauli principle automatically in an antisymmetric spin state (i.e. with the – sign). In contrast, the Coulomb repulsion of the electrons, i.e. the tendency that they try to avoid each other by this repulsion, would lead to an antisymmetric orbital function (i.e. with the – sign) of these two particles, and complementary to a symmetric spin function (i.e. with the + sign, one of the so-called "triplet functions"). Thus, now the spins would be parallel (ferromagnetism in a solid, paramagnetism in two-atomic gases).

The last-mentioned tendency dominates in the metals iron, cobalt and nickel, and in some rare earths, which are *ferromagnetic*. Most of the other metals, where the first-mentioned tendency dominates, are *nonmagnetic* (e.g. sodium, aluminium, and magnesium) or *antiferromagnetic* (e.g. manganese). Diatomic gases are also almost exclusively diamagnetic, and not paramagnetic. However, the oxygen molecule, because of the involvement of π-orbitals, is an exception important for the life-sciences.

The Heitler-London considerations can be generalized to the Heisenberg model of magnetism (Heisenberg 1928).

The explanation of the phenomena is thus essentially based on all subtleties of quantum mechanics, whereas the electrodynamics covers mainly the phenomenology.

Units

SI

SI electromagnetism units			
Symbol	Name of quantity	Derived units	Conversion of international to SI base units
I	Electric current	ampere (SI base unit)	$A=C\,s^{-1}$
q	Electric charge	coulomb	$C=A\,s$
$U, \Delta V, \Delta\phi, E$	Potential difference; Electromotive force	volt	$V=J\,C^{-1}=kg\,A^{-1}m^2s^{-3}$
R; Z; X	Resistance; Impedance; Reactance	ohm	$\Omega=V\,A^{-1}=kg\,m^2\,A^{-2}s^{-3}$

ρ	Resistivity	ohm metre	$\Omega\ m=kg\ A^{-2}m^{3}s^{-3}$
P	Electric power	watt	$W=V\ A=kg\ m^{2}s^{-3}$
C	Capacitance	farad	$F=C\ V^{-1}=A^{2}kg^{-1}m^{-2}s^{4}$
Å	Electric field strength	volt per metre	$V\ m^{-1}=C^{-1}N=kg\ A^{-1}m\ s^{-3}$
D	Electric displacement field	Coulomb per square metre	$C\ m^{-2}=A\ m^{-2}s$
ε	Permittivity	farad per metre	$F\ m^{-1}=A^{2}kg^{-1}m^{-3}s^{4}$
χ_e	Electric susceptibility	Dimensionless	
G; Y; B	Conductance; Admittance; Susceptance	siemens	$S=\Omega^{-1}=kg^{-1}A^{2}m^{-2}s^{3}$
γ, κ, σ	Conductivity	siemens per metre	$S\ m^{-1}=A^{2}kg^{-1}m^{-3}s^{3}$
B	Magnetic flux density, Magnetic induction	tesla	$T=Wb\ m^{-2}=kg\ A^{-1}s^{-2}$
Φ	Magnetic flux	weber	$Wb=V\ s-kg\ A^{-1}m^{2}s^{-2}$
H	Magnetic field strength	ampere per metre	$A\ m^{-1}$
L, M	Inductance	henry	$H=Wb\ A^{-1}=V\ A^{-1}s=kg\ A^{-2}m^{2}s^{-2}$
μ	Permeability	henry per metre	$Hm^{-1}=kg\ A^{-2}m\ s^{-2}$
χ	Magnetic susceptibility	Dimensionless	

Other

- gauss – the centimeter-gram-second (CGS) unit of magnetic field (denoted B)
- oersted – the CGS unit of magnetizing field (denoted H)
- maxwell – the CGS unit for magnetic flux
- gamma – a unit of *magnetic flux density* that was commonly used before the tesla came into use (1.0 gamma = 1.0 nanotesla)
- μ_0 – common symbol for the permeability of free space ($4\pi\times10^{-7}$ newton/(ampere-turn)^2)

Living Things

Some organisms can detect magnetic fields, a phenomenon known as magnetoception. In addition to detection, biomagnetic pheonomena is utilized by organisms in a number of ways. For instance, chitons, a type of marine mollusk, produce magnetite to harden their teeth, and even humans produce magnetite in bodily tissue. Magnetobiology studies magnetic fields as a medical treatment; fields naturally produced by an organism are known as biomagnetism.

References

- Modell, Michael; Robert C. Reid (1974). Thermodynamics and Its Applications. Englewood Cliffs, NJ: Prentice-Hall. ISBN 0-13-914861-2
- Grimm, R. (2005). "Low-temperature physics: A quantum revolution". Nature. 435 (7045): 1035–1036. Bibcode:2005Natur.435.1035G. PMID 15973388. doi:10.1038/4351035a
- Keith Welch. "Questions & Answers - How do you explain electrical resistance?". Thomas Jefferson National Accelerator Facility. Retrieved 28 April 2017
- R. M. Pashley; M. Rzechowicz; L. R. Pashley; M. J. Francis (2005). "De-Gassed Water is a Better Cleaning Agent". The Journal of Physical Chemistry B. 109 (3): 1231–8. PMID 16851085. doi:10.1021/jp045975a
- David Griffiths (1999) [1981]. "7. Electrodynamics". In Alison Reeves. Introduction to Electrodynamics (3rd ed.). Upper Saddle River, New Jersey: Prentice Hall. p. 286. ISBN 0-13-805326-X. OCLC 40251748
- Stephenson, C.; Hubler, A. (2015). "Stability and conductivity of self assembled wires in a transverse electric field". Sci.Rep.5. 8323. doi:10.1038/srep15044
- "CODATA Value: electron g factor". The NIST Reference on Constants, Units, and Uncertainty. NIST. 2006. Retrieved 2013-11-15
- Y. Pauleau, Péter B. Barna, P. B. Barna (1997) Protective coatings and thin films: synthesis, characterization, and applications, p. 215, Springer, ISBN 0-7923-4380-8
- Pauli, Wolfgang (1940). "The Connection Between Spin and Statistics" (PDF). Phys. Rev. 58 (8): 716–722. Bibcode:1940PhRv...58..716P. doi:10.1103/PhysRev.58.716
- Schmidl, Petra G. (1996–1997). "Two Early Arabic Sources On The Magnetic Compass". Journal of Arabic and Islamic Studies. 1: 81–132
- K.F. Riley, M.P. Hobson, S.J. Bence (2010) Mathematical methods for physics and engineering, Cambridge University Press, ISBN 978-0-521-86153-3
- Vowles, Hugh P. (1932). "Early Evolution of Power Engineering". Isis. University of Chicago Press. 17 (2): 412–420 [419–20]. doi:10.1086/346662
- Kirschvink, Joseph L.; Kobayashi-Kirshvink, Atsuko; Diaz-Ricci, Juan C.; Kirschvink, Steven J. (1992). "Magnetite in Human Tissues: A Mechanism for the Biological Effects of Weak ELF Magnetic Fields" (PDF). Bioelectromagnetics Supplement. 1: 101–113. Retrieved 29 March 2016
- Assadi, M.H.N; Hanaor, D.A.H (2013). "Theoretical study on copper's energetics and magnetism in TiO_2 polymorphs" (PDF). Journal of Applied Physics. 113 (23): 233913. Bibcode:2013JAP...113w3913A. arXiv:1304.1854. doi:10.1063/1.4811539
- R.P. Feynman (1985). "Electrons and Their Interactions". QED: The Strange Theory of Light and Matter. Princeton, New Jersey: Princeton University Press. p. 115. ISBN 0-691-08388-6
- A. B. Migdal (1959). "Superfluidity and the moments of inertia of nuclei". Nucl. Phys. 13 (5): 655–674. Bibcode:1959NucPh..13..655M. doi:10.1016/0029-5582(59)90264-0
- Copper wire tables : United States. National Bureau of Standards : Free Download & Streaming : Internet Archive. Archive.org (2001-03-10). Retrieved on 2014-02-03
- Merzouki, Rochdi; Samantaray, Arun Kumar; Pathak, Pushparaj Mani (2012). Intelligent Mechatronic Systems: Modeling, Control and Diagnosis. Springer Science & Business Media. pp. 403–405. ISBN 144714628X

Theoretical Condensed Matter Physics

Phase transition describes the change in state of matter into solid, liquid, gas and rarely, plasma. The physical properties of matter become uniform during phase transitions. Condensed matter theory is best understood in confluence with the major topics listed in the following chapter.

Phase Transition

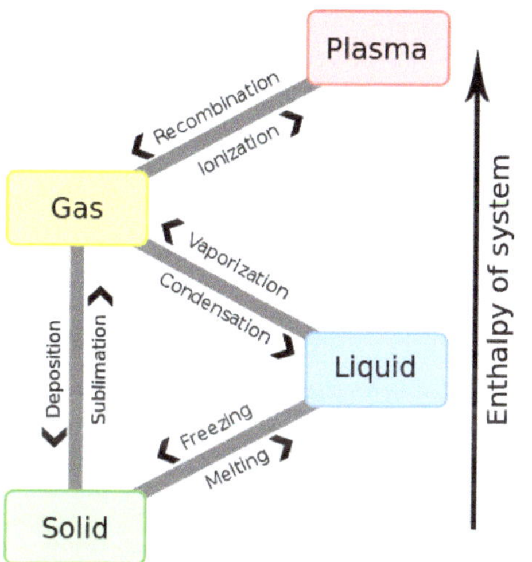

This diagram shows the nomenclature for the different phase transitions

The term phase transition (or phase change) is most commonly used to describe transitions between solid, liquid and gaseous states of matter, and, in rare cases, plasma (physics). A phase of a thermodynamic system and the states of matter have uniform physical properties. During a phase transition of a given medium certain properties of the medium change, often discontinuously, as a result of the change of some external condition, such as temperature, pressure, or others. For example, a liquid may become gas upon heating to the boiling point, resulting in an abrupt change in volume. The measurement of the external conditions at which the transformation occurs is termed the phase transition. Phase transitions are common in nature and used today in many technologies.

Types of Phase Transition

Examples of phase transitions include:

- The transitions between the solid, liquid, and gaseous phases of a single component, due to the effects of temperature and/or pressure:

Solid		To			
		Liquid	**Gas**	**Plasma**	
From	Solid	Solid-solid transformation	Melting	Sublimation	—
	Liquid	Freezing	—	Boiling/evaporation	—
	Gas	Deposition	Condensation	—	Ionization
	Plasma	—	—	Recombination/deionization	—

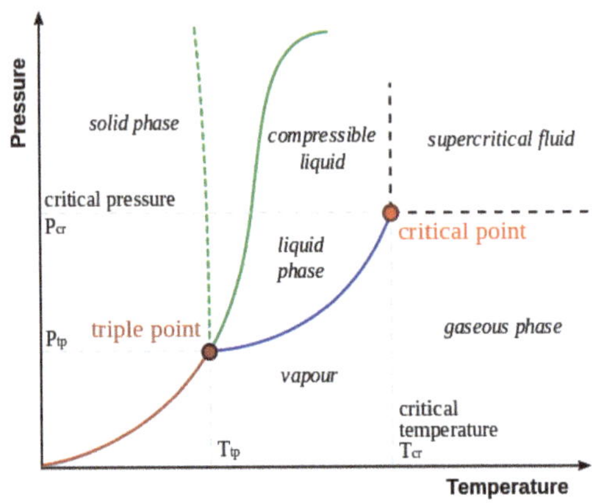

A typical phase diagram. The dotted line gives the anomalous behavior of water

A small piece of rapidly melting solid argon simultaneously shows the transitions from solid to liquid and liquid to gas

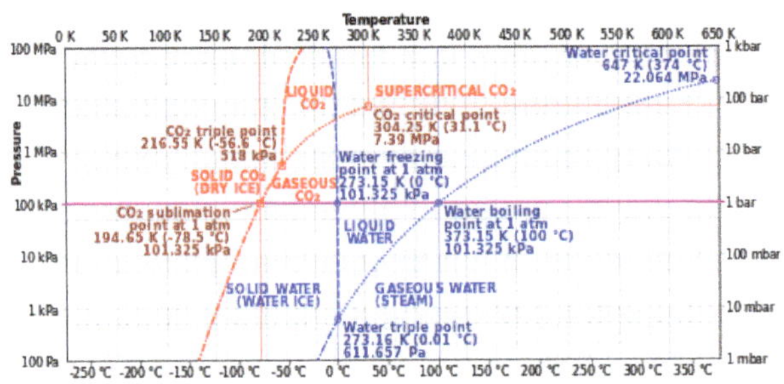

Comparison of phase diagrams of carbon dioxide (red) and water (blue) explaining their different phase transitions at 1 atmosphere.

- A eutectic transformation, in which a two component single phase liquid is cooled and transforms into two solid phases. The same process, but beginning with a solid instead of a liquid is called a eutectoid transformation.
- A peritectic transformation, in which a two component single phase solid is heated and transforms into a solid phase and a liquid phase.
- A spinodal decomposition, in which a single phase is cooled and separates into two different compositions of that same phase.
- Transition to a mesophase between solid and liquid, such as one of the "liquid crystal" phases.
- The transition between the ferromagnetic and paramagnetic phases of magnetic materials at the Curie point.
- The transition between differently ordered, commensurate or incommensurate, magnetic structures, such as in cerium antimonide.
- The martensitic transformation which occurs as one of the many phase transformations in carbon steel and stands as a model for displacive phase transformations.
- Changes in the crystallographic structure such as between ferrite and austenite of iron.
- Order-disorder transitions such as in alpha-titanium aluminides.
- The dependence of the adsorption geometry on coverage and temperature, such as for hydrogen on iron (110).
- The emergence of superconductivity in certain metals and ceramics when cooled below a critical temperature.
- The transition between different molecular structures (polymorphs, allotropes or polyamorphs), especially of solids, such as between an amorphous structure and a crystal structure, between two different crystal structures, or between two amorphous structures.
- Quantum condensation of bosonic fluids (Bose–Einstein condensation). The superfluid transition in liquid helium is an example of this.
- The breaking of symmetries in the laws of physics during the early history of the universe as its temperature cooled.
- Isotope fractionation occurs during a phase transition, the ratio of light to heavy isotopes in the involved molecules changes. When water vapor condenses (an equilibrium fractionation), the heavier water isotopes (18O and 2H) become enriched in the liquid phase while the lighter isotopes (16O and 1H) tend toward the vapor phase.

Phase transitions occur when the thermodynamic free energy of a system is non-analytic for some choice of thermodynamic variables (cf. phases). This condition generally stems from the interactions of a large number of particles in a system, and does not appear in systems that are too small. It is important to note that phase transitions can occur and are defined for non-thermodynamic systems, where temperature is not a parameter. Examples include: quantum phase transitions, dynamic phase transitions, and topological (structural) phase transitions. In these types of systems other parameters take the place of temperature. For instance, connection probability replaces temperature for percolating networks.

At the phase transition point (for instance, boiling point) the two phases of a substance, liquid and vapor, have identical free energies and therefore are equally likely to exist. Below the boiling point, the liquid is the more stable state of the two, whereas above the gaseous form is preferred.

It is sometimes possible to change the state of a system diabatically (as opposed to adiabatically) in such a way that it can be brought past a phase transition point without undergoing a phase transition. The resulting state is metastable, i.e., less stable than the phase to which the transition would have occurred, but not unstable either. This occurs in superheating, supercooling, and supersaturation, for example.

Classifications

Ehrenfest Classification

Paul Ehrenfest classified phase transitions based on the behavior of the thermodynamic free energy as a function of other thermodynamic variables. Under this scheme, phase transitions were labeled by the lowest derivative of the free energy that is discontinuous at the transition. *First-order phase transitions* exhibit a discontinuity in the first derivative of the free energy with respect to some thermodynamic variable. The various solid/liquid/gas transitions are classified as first-order transitions because they involve a discontinuous change in density, which is the (inverse of the) first derivative of the free energy with respect to pressure. *Second-order phase transitions* are continuous in the first derivative (the order parameter, which is the first derivative of the free energy with respect to the external field, is continuous across the transition) but exhibit discontinuity in a second derivative of the free energy. These include the ferromagnetic phase transition in materials such as iron, where the magnetization, which is the first derivative of the free energy with respect to the applied magnetic field strength, increases continuously from zero as the temperature is lowered below the Curie temperature. The magnetic susceptibility, the second derivative of the free energy with the field, changes discontinuously. Under the Ehrenfest classification scheme, there could in principle be third, fourth, and higher-order phase transitions.

Though useful, Ehrenfest's classification has been found to be an incomplete method of classifying phase transitions, for it does not take into account the case where a derivative of free energy diverges (which is only possible in the thermodynamic limit). For instance, in the ferromagnetic transition, the heat capacity diverges to infinity. The same phenomenon is also seen in superconducting phase transition.

Modern Classifications

In the modern classification scheme, phase transitions are divided into two broad categories, named similarly to the Ehrenfest classes:

First-order phase transitions are those that involve a latent heat. During such a transition, a system either absorbs or releases a fixed (and typically large) amount of energy per volume. During this process, the temperature of the system will stay constant as heat is added: the system is in a "mixed-phase regime" in which some parts of the system have completed the transition and others have not. Familiar examples are the melting of ice or the boiling of water (the water does not instantly turn into vapor, but forms a turbulent mixture of liquid water and vapor bubbles). Imry and Wortis showed that quenched disorder can broaden a first-order transition in that the transformation is completed over a finite range of temperatures, but phenomena like supercooling and superheating survive and hysteresis is observed on thermal cycling.

Second-order phase transitions are also called *continuous phase transitions*. They are characterized by a divergent susceptibility, an infinite correlation length, and a power-law decay of correlations near criticality. Examples of second-order phase transitions are the ferromagnetic transition, superconducting transition (for a Type-I superconductor the phase transition is second-order at zero external field and for a Type-II superconductor the phase transition is second-order for both normal-state--mixed-state and mixed-state--superconducting-state transitions) and the superfluid transition. In contrast to viscosity, thermal expansion and heat capacity of amorphous materials show a relatively sudden change at the glass transition temperature which enables accurate detection using differential scanning calorimetry measurements. Lev Landau gave a phenomenological theory of second-order phase transitions.

Apart from isolated, simple phase transitions, there exist transition lines as well as multicritical points, when varying external parameters like the magnetic field or composition.

Several transitions are known as the *infinite-order phase transitions*. They are continuous but break no symmetries. The most famous example is the Kosterlitz–Thouless transition in the two-dimensional XY model. Many quantum phase transitions, e.g., in two-dimensional electron gases, belong to this class.

The liquid–glass transition is observed in many polymers and other liquids that can be supercooled far below the melting point of the crystalline phase. This is atypical in several respects. It is not a transition between thermodynamic ground states: it is widely believed that the true ground state is always crystalline. Glass is a *quenched disorder* state, and its entropy, density, and so on, depend on the thermal history. Therefore, the glass transition is primarily a dynamic phenomenon: on cooling a liquid, internal degrees of freedom successively fall out of equilibrium. Some theoretical methods predict an underlying phase transition in the hypothetical limit of infinitely long relaxation times. No direct experimental evidence supports the existence of these transitions.

Characteristic Properties

Phase Coexistence

A disorder-broadened first-order transition occurs over a finite range of temperatures where the fraction of the low-temperature equilibrium phase grows from zero to one (100%) as the temperature is lowered. This continuous variation of the coexisting fractions with temperature raised interesting possibilities. On cooling, some liquids vitrify into a glass rather than transform to the equilibrium crystal phase. This happens if the cooling rate is faster than a critical cooling rate, and is attributed to the molecular motions becoming so slow that the molecules cannot rearrange into the crystal positions. This slowing down happens below a glass-formation temperature Tg, which may depend on the applied pressure., If the first-order freezing transition occurs over a range of temperatures, and Tg falls within this range, then there is an interesting possibility that the transition is arrested when it is partial and incomplete. Extending these ideas to first-order magnetic transitions being arrested at low temperatures, resulted in the observation of incomplete magnetic transitions, with two magnetic phases coexisting, down to the lowest temperature. First reported in the case of a ferromagnetic to anti-ferromagnetic transition, such persistent phase coexistence has now been reported across a variety of first-order magnetic transitions. These include colossal-magnetoresistance manganite materials, magnetocaloric materials, magnetic shape memory materials, and other materials. The interesting feature of these observations of Tg falling within the temperature range over which the transition occurs is that the first-order magnetic transition is influenced by magnetic field, just like the structural transition is influenced by pressure. The relative ease with which magnetic field can be controlled, in contrast to pressure, raises the possibility that one can study the interplay between Tg and Tc in an exhaustive way. Phase coexistence across first-order magnetic transitions will then enable the resolution of outstanding issues in understanding glasses.

Critical Points

In any system containing liquid and gaseous phases, there exists a special combination of pressure and temperature, known as the critical point, at which the transition between liquid and gas becomes a second-order transition. Near the critical point, the fluid is sufficiently hot and compressed that the distinction between the liquid and gaseous phases is almost non-existent. This is associated with the phenomenon of critical opalescence, a milky appearance of the liquid due to density fluctuations at all possible wavelengths (including those of visible light).

Symmetry

Phase transitions often involve a symmetry breaking process. For instance, the cooling of a fluid into a crystalline solid breaks continuous translation symmetry: each point in the fluid has the same properties, but each point in a crystal does not have the same properties (unless the points are chosen from the lattice points of the crystal lattice). Typically, the high-temperature phase contains more symmetries than the low-temperature phase due to spontaneous symmetry breaking, with the exception of certain accidental symmetries (e.g. the formation of heavy virtual particles, which only occurs at low temperatures).

Order Parameters

An order parameter is a measure of the degree of order across the boundaries in a phase transition system; it normally ranges between zero in one phase (usually above the critical point) and nonzero in the other. At the critical point, the order parameter susceptibility will usually diverge.

An example of an order parameter is the net magnetization in a ferromagnetic system undergoing a phase transition. For liquid/gas transitions, the order parameter is the difference of the densities.

From a theoretical perspective, order parameters arise from symmetry breaking. When this happens, one needs to introduce one or more extra variables to describe the state of the system. For example, in the ferromagnetic phase, one must provide the net magnetization, whose direction was spontaneously chosen when the system cooled below the Curie point. However, note that order parameters can also be defined for non-symmetry-breaking transitions. Some phase transitions, such as superconducting and ferromagnetic, can have order parameters for more than one degree of freedom. In such phases, the order parameter may take the form of a complex number, a vector, or even a tensor, the magnitude of which goes to zero at the phase transition.

There also exist dual descriptions of phase transitions in terms of disorder parameters. These indicate the presence of line-like excitations such as vortex- or defect lines.

Relevance in Cosmology

Symmetry-breaking phase transitions play an important role in cosmology. It has been speculated by Lee Smolin and Benjamin and Jeremy Bernstein that, in the hot early universe, the vacuum (i.e. the various quantum fields that fill space) possessed a large number of symmetries. As the universe expanded and cooled, the vacuum underwent a series of symmetry-breaking phase transitions. For example, the electroweak transition broke the SU(2)×U(1) symmetry of the electroweak field into the U(1) symmetry of the present-day electromagnetic field. This transition is important to understanding the asymmetry between the amount of matter and antimatter in the present-day universe.

Progressive phase transitions in an expanding universe are implicated in the development of order in the universe, as is illustrated by the work of Eric Chaisson and David Layzer.

Critical Exponents and Universality Classes

Continuous phase transitions are easier to study than first-order transitions due to the absence of latent heat, and they have been discovered to have many interesting properties. The phenomena associated with continuous phase transitions are called critical phenomena, due to their association with critical points.

It turns out that continuous phase transitions can be characterized by parameters known as critical exponents. The most important one is perhaps the exponent describing the divergence of the thermal correlation length by approaching the transition. For instance, let us examine the behavior of the heat capacity near such a transition. We vary the temperature T of the system while keeping all the other thermodynamic variables fixed, and find that the transition occurs at some critical temperature T_c. When T is near T_c, the heat capacity C typically has a power law behavior,

$$C \propto |T_c - T|^{-\alpha}.$$

The heat capacity of amorphous materials has such a behaviour near the glass transition temperature where the universal critical exponent α = 0.59 A similar behavior, but with the exponent ν instead of α, applies for the correlation length.

The exponent ν is positive. This is different with α. Its actual value depends on the type of phase transition we are considering.

It is widely believed that the critical exponents are the same above and below the critical temperature. It has now been shown that this is not necessarily true: When a continuous symmetry is explicitly broken down to a discrete symmetry by irrelevant (in the renormalization group sense) anisotropies, then some exponents (such as γ, the exponent of the susceptibility) are not identical.

For $-1 < \alpha < 0$, the heat capacity has a "kink" at the transition temperature. This is the behavior of liquid helium at the lambda transition from a normal state to the superfluid state, for which experiments have found $\alpha = -0.013\pm0.003$. At least one experiment was performed in the zero-gravity conditions of an orbiting satellite to minimize pressure differences in the sample. This experimental value of α agrees with theoretical predictions based on variational perturbation theory.

For $0 < \alpha < 1$, the heat capacity diverges at the transition temperature (though, since $\alpha < 1$, the enthalpy stays finite). An example of such behavior is the 3D ferromagnetic phase transition. In the three-dimensional Ising model for uniaxial magnets, detailed theoretical studies have yielded the exponent $\alpha \sim +0.110$.

Some model systems do not obey a power-law behavior. For example, mean field theory predicts a finite discontinuity of the heat capacity at the transition temperature, and the two-dimensional Ising model has a logarithmic divergence. However, these systems are limiting cases and an exception to the rule. Real phase transitions exhibit power-law behavior.

Several other critical exponents, β, γ, δ, ν, and η, are defined, examining the power law behavior of a measurable physical quantity near the phase transition. Exponents are related by scaling relations, such as

$$\beta = \gamma / (\delta - 1), \qquad \nu = \gamma / (2 - \eta).$$

It can be shown that there are only two independent exponents, e.g. ν and η.

It is a remarkable fact that phase transitions arising in different systems often possess the same set of critical exponents. This phenomenon is known as *universality*. For example, the critical exponents at the liquid–gas critical point have been found to be independent of the chemical composition of the fluid.

More impressively, but understandably from above, they are an exact match for the critical exponents of the ferromagnetic phase transition in uniaxial magnets. Such systems are said to be in the same universality class. Universality is a prediction of the renormalization group theory of phase transitions, which states that the thermodynamic properties of a system near a phase transition depend only on a small number of features, such as dimensionality and symmetry, and are insensitive to the underlying microscopic properties of the system. Again, the divergence of the correlation length is the essential point.

Critical Slowing Down and Other Phenomena

There are also other critical phenomena; e.g., besides *static functions* there is also *critical dynamics*. As a consequence, at a phase transition one may observe critical slowing down or *speeding up*. The large *static universality classes* of a continuous phase transition split into smaller *dynamic universality* classes. In addition to the critical exponents, there are also universal relations for certain static or dynamic functions of the magnetic fields and temperature differences from the critical value.

Percolation Theory

Another phenomenon which shows phase transitions and critical exponents is percolation. The simplest example is perhaps percolation in a two dimensional square lattice. Sites are randomly occupied with probability p. For small values of p the occupied sites form only small clusters. At a certain threshold p_c a giant cluster is formed and we have a second-order phase transition. The behavior of P_∞ near p_c is, $P_\infty \sim (p-p_c)^\beta$, where β is a critical exponent.

Phase Transitions in Biological Systems

Phase transitions play many important roles in biological systems. Examples include the lipid bilayer formation, the coil-globule transition in the process of protein folding and DNA melting, liquid crystal-like transitions in the process of DNA condensation, and cooperative ligand binding to DNA and proteins with the character of phase transition.

In *biological membranes*, gel to liquid crystalline phase transitions play a critical role in physiological functioning of biomembranes. In gel phase, due to low fluidity of membrane lipid fatty-acyl chains, membrane proteins have restricted movement and thus are restrained in exercise of their physiological role. Plants depend critically on photosynthesis by chloroplast thylakoid membranes which are exposed cold environmental temperatures. Thylakoid membranes retain innate fluidity even at relatively low temperatures because of high degree of fatty-acyl disorder allowed by their high content of linolenic acid, 18-carbon chain with 3-double bonds. Gel-to-liquid crystalline phase transition temperature of biological membranes can be determined by many techniques including calorimetry, flouorescence, spin label electron paramagnetic resonance and NMR by recording measurements of the concerned parameter by at series of sample temperatures. A simple method for its determination from 13-C NMR line intensities has also been proposed.

It has been proposed that some biological systems might lie near critical points. Examples include neural networks in the salamander retina, bird flocks gene expression networks in Drosophila, and protein folding. However, it is not clear whether or not alternative reasons could explain some of the phenomena supporting arguments for criticality. It has also been suggested that biological organisms share two key properties of phase transitions: the change of macroscopic behavior and the coherence of a system at a critical point.

Critical Exponent

Critical exponents describe the behavior of physical quantities near continuous phase transitions. It is believed, though not proven, that they are universal, i.e. they do not depend on the details of the physical system, but only on

- the dimension of the system,
- the range of the interaction,
- the spin dimension.

These properties of critical exponents are supported by experimental data. The experimental results can be theoretically achieved in mean field theory for higher-dimensional systems (4 or more dimensions). The theoretical treatment of lower-dimensional systems (1 or 2 dimensions) is more difficult and requires the renormalization group. Phase transitions and critical exponents appear also in percolation systems. However, here the critical dimension above which mean field exponents are valid is 6 and higher dimensions. Mean field critical exponents are also valid for random graphs, such as Erdős–Rényi graphs, which can be regarded as infinite dimensional systems.

Definition

Phase transitions occur at a certain temperature, called the critical temperature T_c. We want to describe the behavior of a physical quantity in terms of a power law around the critical temperature. So we introduce the reduced temperature $\tau := (T - T_c)/T_c$, which is zero at the phase transition, and define the critical exponent k.

$$k \overset{def}{=} \lim_{\tau \to 0} \frac{\log |f(\tau)|}{\log |\tau|}.$$

This results in the power law we were looking for.

$$f(\tau) \propto \tau^k, \quad \tau \approx 0.$$

It is important to remember that this represents the asymptotic behavior of the function $f(\tau)$ as $\tau \to 0$.

More generally one might expect

$$f(\tau) = A\tau^k (1 + b\tau^{k_1} + \cdots).$$

The Most Important Critical Exponents

Below T_c the system has two different phases characterized by an order parameter Ψ, which vanishes at and above T_c.

Let us consider the disordered phase (τ > 0), ordered phase (τ < 0) and critical temperature (τ = 0) phases separately. Following the standard convention, the critical exponents related to the ordered phase are primed. It's also another standard convention to use the super/subscript +(-) for the disordered(ordered) state. We have spontaneous symmetry breaking in the ordered phase. So, we will arbitrarily take any solution in the phase.

Keys

Ψ	order parameter (e.g. $\frac{\rho - \rho_c}{\rho_c}$ for the liquid-gas critical point, magnetization for the Curie point,etc.)
τ	$\frac{T - T_c}{T_c}$
f	specific free energy

C	specific heat; $-T\frac{\partial^2 f}{\partial T^2}$
J	source field (e.g. $\frac{P-P_c}{P_c}$ where P is the pressure and Pc the critical pressure for the liquid-gas critical point, reduced chemical potential, the magnetic field H for the Curie point)
χ	the susceptibility/compressibility/etc.; $\frac{\partial \Psi}{\partial J}$
ξ	correlation length
d	the number of spatial dimensions
$\langle \psi(\vec{x})\psi(\vec{y})\rangle$	the correlation function
r	spatial distance

The following entries are evaluated at $J=0$ (except for the δ entry)

Critical exponents for τ > 0 (disordered phase)	
Greek letter	relation
α	$C \propto \tau^{-\alpha}$
γ	$\chi \propto \tau^{-\gamma}$
ν	$\xi \propto \tau^{-\nu}$
Critical exponents for τ < 0 (ordered phase)	
Greek letter	relation
α'	$C \propto (-\tau)^{-\alpha'}$
β	$\Psi \propto (-\tau)^{\beta}$
γ'	$\chi \propto (-\tau)^{-\gamma'}$
ν'	$\xi \propto (-\tau)^{-\nu'}$
Critical exponents for τ = 0	
Greek letter	relation
δ	$J \propto \Psi^{\delta}$
η	$\langle \psi(0)\psi(r)\rangle \propto r^{-d+2-\eta}$

The critical exponents can be derived from the specific free energy $f(J,T)$ as a function of the source and temperature. The correlation length can be derived from the functional $F[J;T]$.

These relations are accurate close to the critical point in two- and three-dimensional systems. In four dimensions, however, the power laws are modified by logarithmic factors. This problem does not appear in 3.99 dimensions, though.

Mean Field Critical Exponents of Ising-like Systems

The classical Landau theory (aka mean field theory) values of the critical exponents for a scalar field (of which the Ising-model is the prototype example) are given by

$$\alpha = \alpha' = 0$$

$$\beta = \frac{1}{2}$$

$$\gamma = \gamma' = 1$$

$$\delta = 3$$

If we add derivative terms turning it into a mean field Ginzburg–Landau theory, we get

$$\eta = 0$$

$$\nu = \frac{1}{2}$$

One of the major discoveries in the study of critical phenomena is that mean field theory of critical points is only correct when the space dimension of the system is four or higher (which unfortunately excludes many of the experimentally relevant cases). This dimension is called the upper critical dimension. The problem with mean field theory is that the critical exponents do not depend on the space dimension. This leads to a quantitative discrepancy in space dimensions 2 and 3, where the true critical exponents differ from the mean field values. It leads to a qualitative discrepancy in space dimension 1, where a critical point in fact no longer exists, even though mean field theory still predicts there is one. The space dimension where mean field theory becomes qualitatively incorrect is called the lower critical dimension.

Experimental Values

The most accurately measured value of α is −0.0127(3) for the phase transition of superfluid helium (the so-called lambda transition). The value was measured on a space shuttle to minimize pressure differences in the sample. Interestingly, this value is in a significant disagreement with the most precise theoretical determination by a combination of Monte Carlo and high temperature expansion techniques. Other techniques give results in agreement in the experiment but are less precise.

Scaling Functions

In light of the critical scalings, we can reexpress all thermodynamic quantities in terms of dimensionless quantities. Close enough to the critical point, everything can be reexpressed in terms of certain ratios of the powers of the reduced quantities. These are the scaling functions.

The origin of scaling functions can be seen from the renormalization group. The critical point is an infrared fixed point. In a sufficiently small neighborhood of the critical point, we may linearize the action of the renormalization group. This basically means that rescaling the system by a factor of a will be equivalent to rescaling operators and source fields by a factor of a^{Δ} for some Δ. So, we may reparameterize all quantities in terms of rescaled scale independent quantities.

Scaling Relations

It has been believed for a long time that the critical exponents were the same above and below the critical temperature, e.g. : $\alpha \equiv \alpha'$ or $\gamma \equiv \gamma'$. It has now been shown that this is not necessarily true: When a continuous symmetry is explicitly broken down to a discrete symmetry by irrelevant (in the renormalization group sense) anisotropies, then the exponents γ and γ' are not identical.

Critical exponents are denoted by Greek letters. They fall into universality classes and obey the scaling relations

$$\nu d = 2 - \alpha = 2\beta + \gamma = \beta(\delta + 1) = \gamma \frac{\delta + 1}{\delta - 1}$$

$$2 - \eta = \frac{\gamma}{\nu} = d \frac{\delta - 1}{\delta + 1}$$

These equations imply that there are only two independent exponents, e.g., ν and η . All this follows from the theory of the renormalization group.

Anisotropy

There are some anisotropic systems where the correlation length is direction dependent.

Multicritical Points

More complex behavior may occur at multicritical points, at the border or on intersections of critical manifolds.

Static Versus Dynamic Properties

The above examples exclusively refer to the static properties of a critical system. However dynamic properties of the system may become critical, too. Especially, the characteristic time, τ_{char}, of a system diverges as $\tau_{char} \propto \xi^{z}$, with a dynamical exponent z . Moreover, the large static universality classes of equivalent models with identical static critical exponents decompose into smaller dynamical universality classes, if one demands that also the dynamical exponents are identical.

The critical exponents can be computed from conformal field theory.

Transport Properties

Critical exponents also exist for transport quantities like viscosity and heat conductivity.

Self-organized criticality

Critical exponents also exist for self organized criticality for dissipative systems.

Percolation Theory

Phase transitions and critical exponents appear also in percolation processes where the concentration of occupied sites or links play the role of temperature.

Emergence

The formation of complex symmetrical and fractal patterns in snowflakes exemplifies emergence in a physical system.

A termite “cathedral” mound produced by a termite colony offers a classic example of emergence in nature.

In philosophy, systems theory, science, and art, emergence is a phenomenon whereby larger entities arise through interactions among smaller or simpler entities such that the larger entities exhibit properties the smaller/simpler entities do not exhibit.

Emergence is central in theories of integrative levels and of complex systems. For instance, the phenomenon of *life* as studied in biology is an emergent property of chemistry, and psychological phenomena emerge from the neurobiological phenomena of living things.

In philosophy, theories that emphasize emergent properties have been called emergentism. Almost all accounts of emergentism include a form of epistemic or ontological irreducibility to the lower levels.

In Philosophy

In philosophy, emergence is often understood to be a claim about the etiology of a system's properties. An emergent property of a system, in this context, is one that is not a property of any component of that system, but is still a feature of the system as a whole. Nicolai Hartmann, one of the first modern philosophers to write on emergence, termed this *categorial novum* (new category).

Definitions

This idea of emergence has been around since at least the time of Aristotle. John Stuart Mill and Julian Huxley are two of many scientists and philosophers who have written on the concept.

The term "emergent" was coined by philosopher G. H. Lewes, who wrote:

Every resultant is either a sum or a difference of the co-operant forces; their sum, when their directions are the same – their difference, when their directions are contrary. Further, every resultant is clearly traceable in its components, because these are homogeneous and commensurable. It is otherwise with emergents, when, instead of adding measurable motion to measurable motion, or things of one kind to other individuals of their kind, there is a co-operation of things of unlike kinds. The emergent is unlike its components insofar as these are incommensurable, and it cannot be reduced to their sum or their difference.

Economist Jeffrey Goldstein provided a current definition of emergence in the journal *Emergence*. Goldstein initially defined emergence as: "the arising of novel and coherent structures, patterns and properties during the process of self-organization in complex systems".

Goldstein's definition can be further elaborated to describe the qualities of this definition in more detail:

The common characteristics are: (1) radical novelty (features not previously observed in systems); (2) coherence or correlation (meaning integrated wholes that maintain themselves over some period of time); (3) A global or macro "level" (i.e. there is some property of "wholeness"); (4) it is the product of a dynamical process (it evolves); and (5) it is "ostensive" (it can be perceived).

Systems scientist Peter Corning also says that living systems cannot be reduced to underlying laws of physics:

Rules, or laws, have no causal efficacy; they do not in fact "generate" anything. They serve merely to describe regularities and consistent relationships in nature. These patterns may be very illuminating and important, but the underlying causal agencies must be separately specified (though often they are not). But that aside, the game of chess illustrates ... why any laws or rules of emergence and evolution are insufficient. Even in a chess game, you cannot use the rules to predict "history" – i.e., the course of any given game. Indeed, you cannot even reliably predict the next move in a chess game. Why? Because the "system" involves more than the rules of the game. It also includes the players and their unfolding, moment-by-moment decisions among a very large number of available options at each choice point. The game of chess is inescapably historical, even though it is also constrained and shaped by a set of rules, not to mention the laws of physics. Moreover, and this is a key point, the game of chess is also shaped by teleonomic, cybernetic, feedback-driven influences. It is not simply a self-ordered process; it involves an organized, "purposeful" activity.

Strong and Weak Emergence

Usage of the notion "emergence" may generally be subdivided into two perspectives, that of "weak emergence" and "strong emergence". In terms of physical systems, weak emergence is a type of emergence in which the emergent property is amenable to computer simulation. This is opposed to the older notion of strong emergence, in which the emergent property cannot be simulated by a computer.

Some common points between the two notions are that emergence concerns new properties produced as the system grows, which is to say ones which are not shared with its components or prior states. Also, it is assumed that the properties are supervenient rather than metaphysically primitive (Bedau 1997).

Weak emergence describes new properties arising in systems as a result of the interactions at an elemental level. However, it is stipulated that the properties can be determined by observing or simulating the system, and not by any process of a priori analysis.

Bedau notes that weak emergence is not a universal metaphysical solvent, as weak emergence leads to the conclusion that matter itself contains elements of awareness to it. However, Bedau concludes that adopting this view would provide a precise notion that emergence is involved in consciousness, and second, the notion of weak emergence is metaphysically benign.(Bedau 1997).

Strong emergence describes the direct causal action of a high-level system upon its components; qualities produced this way are irreducible to the system's constituent parts (Laughlin 2005). The whole is other than the sum of its parts. An example from physics of such emergence is water, being seemingly unpredictable even after an exhaustive study of the properties of its constituent atoms of hydrogen and oxygen. It follows then that no simulation of the system can exist, for such a simulation would itself constitute a reduction of the system to its constituent parts. (Bedau 1997).

However, "the debate about whether or not the whole can be predicted from the properties of the parts misses the point. Wholes produce unique combined effects, but many of these effects may be co-determined by the context and the interactions between the whole and its environment(s)" (Corning 2002). In accordance with his Synergism Hypothesis, (Corning 1983 2005) Corning also stated, "It is the synergistic effects produced by wholes that are the very cause of the evolution of complexity in nature." Novelist Arthur Koestler used the metaphor of Janus (a symbol of the unity underlying complements like open/shut, peace/war) to illustrate how the two perspectives (strong vs. weak or holistic vs. reductionistic) should be treated as non-exclusive, and should work together to address the issues of emergence. (Koestler 1969) Further,

The ability to reduce everything to simple fundamental laws does not imply the ability to start from those laws and reconstruct the universe. The constructionist hypothesis breaks down when confronted with the twin difficulties of scale and complexity. At each level of complexity entirely new properties appear. Psychology is not applied biology, nor is biology applied chemistry. We can now see that the whole becomes not merely more, but very different from the sum of its parts. (Anderson 1972).

The plausibility of strong emergence is questioned by some as contravening our usual understanding of physics. Mark A. Bedau observes:

Although strong emergence is logically possible, it is uncomfortably like magic. How does an irreducible but supervenient downward causal power arise, since by definition it cannot be due to the aggregation of the micro-level potentialities? Such causal powers would be quite unlike anything within our scientific ken. This not only indicates how they will discomfort reasonable forms of materialism. Their mysteriousness will only heighten the traditional worry that emergence entails illegitimately getting something from nothing.

Strong emergence can be criticized for being causally overdetermined. The canonical example concerns emergent mental states (M and M∗) that supervene on physical states (P and P∗) respectively. Let M and M∗ be emergent properties. Let M∗ supervene on base property P∗. What happens when M causes M∗? Jaegwon Kim says:

In our schematic example above, we concluded that M causes M∗ by causing P∗. So M causes P∗. Now, M, as an emergent, must itself have an emergence base property, say P. Now we face a critical question: if an emergent, M, emerges from basal condition P, why cannot P displace M as a cause of any putative effect of M? Why cannot P do all the work in explaining why any alleged effect of M occurred? If causation is understood as nomological (law-based) sufficiency, P, as M's emergence base, is nomologically sufficient for it, and M, as P∗'s cause, is nomologically sufficient for P∗. It follows that P is nomologically sufficient for P∗ and hence qualifies as its cause...If M is somehow retained as a cause, we are faced with the highly implausible consequence that every case of downward causation involves overdetermination (since P remains a cause of P∗ as well). Moreover, this goes against the spirit of emergentism in any case: emergents are supposed to make distinctive and novel causal contributions.

If M is the cause of M∗, then M∗ is overdetermined because M∗ can also be thought of as being determined by P. One escape route that a strong emergentist could take would be to deny downward causation. However, this would deny that emergent mental states must supervene on physical states, which in turn would deny physicalism, and thus be unpalatable for some philosophers and physicists.

Meanwhile, others have worked towards developing analytical evidence of strong emergence. In 2009, Gu et al. presented a class of physical systems that exhibits non-computable macroscopic properties. More precisely, if one could compute certain macroscopic properties of these systems from the microscopic description of these systems, then one would be able to solve computational problems known to be undecidable in computer science. They concluded that Although macroscopic concepts are essential for understanding our world, much of fundamental physics has been devoted to the search for a 'theory of everything', a set of equations that perfectly describe the behavior of all fundamental particles. The view that this is the goal of science rests in part on the rationale that such a theory would allow us to derive the behavior of all macroscopic concepts, at least in principle. The evidence we have presented suggests that this view may be overly optimistic. A `theory of everything' is one of many components necessary for complete understanding of the universe, but is not necessarily the only one. The development of macroscopic laws from first principles may involve more than just systematic logic, and could require conjectures suggested by experiments, simulations or insight.

Emergent structures are patterns that emerge via collective actions of many individual entities. To explain such patterns, one might conclude, per Aristotle, that emergent structures are other than the sum of their parts on the assumption that the emergent order will not arise if the various parts simply interact independently of one another. However, there are those who disagree. According to this argument, the interaction of each part with its immediate surroundings causes a complex chain of processes that can lead to order in some form. In fact, some systems in nature are observed to exhibit emergence based upon the interactions of autonomous parts, and some others exhibit emergence that at least at present cannot be reduced in this way. In particular renormalization are methods in theoretical physics which enables scientists to study systems that are not tractable as the combination of their parts.

Objective or Subjective Quality

The properties of complexity and organization of any system are considered by Crutchfield to be subjective qualities determined by the observer.

Defining structure and detecting the emergence of complexity in nature are inherently subjective, though essential, scientific activities. Despite the difficulties, these problems can be analysed in terms of how model-building observers infer from measurements the computational capabilities embedded in non-linear processes. An observer's notion of what is ordered, what is random, and what is complex in its environment depends directly on its computational resources: the amount of raw measurement data, of memory, and of time available for estimation and inference. The discovery of structure in an environment depends more critically and subtly, though, on how those resources are organized. The descriptive power of the observer's chosen (or implicit) computational model class, for example, can be an overwhelming determinant in finding regularity in data. (Crutchfield 1994).

On the other hand, Peter Corning argues "Must the synergies be perceived/observed in order to qualify as emergent effects, as some theorists claim? Most emphatically not. The synergies associated with emergence are real and measurable, even if nobody is there to observe them."(Corning 2002).

In Religion, Art and Humanities

In religion, emergence grounds expressions of religious naturalism and syntheism in which a sense of the sacred is perceived in the workings of entirely naturalistic processes by which more complex forms arise or evolve from simpler forms. Examples are detailed in *The Sacred Emergence of Nature* by Ursula Goodenough & Terrence Deacon and *Beyond Reductionism: Reinventing the Sacred* by Stuart Kauffman, both from 2006, and in *Syntheism – Creating God in The Internet Age* by Alexander Bard & Jan Söderqvist from 2014. An early argument (1904–05) for the emergence of social formations, in part stemming from religion, can be found in Max Weber's most famous work, *The Protestant Ethic and the Spirit of Capitalism.*

In art, emergence is used to explore the origins of novelty, creativity, and authorship. Some art/literary theorists (Wheeler, 2006; Alexander, 2011 have proposed alternatives to postmodern understandings of "authorship" using the complexity sciences and emergence theory. They contend that artistic selfhood and meaning are emergent, relatively objective phenomena. Michael J. Pearce has used emergence to describe the experience of works of art in relation to contemporary neuroscience).

In international development, concepts of emergence have been used within a theory of social change termed SEED-SCALE to show how standard principles interact to bring forward socio-economic development fitted to cultural values, community economics, and natural environment (local solutions emerging from the larger socio-econo-biosphere). These principles can be implemented utilizing a sequence of standardized tasks that self-assemble in individually specific ways utilizing recursive evaluative criteria.

In postcolonial studies, the term "Emerging Literature" refers to a contemporary body of texts that is gaining momentum in the global literary landscape (v. esp.: J.M. Grassin, ed. *Emerging Literatures*, Bern, Berlin, etc. : Peter Lang, 1996). By opposition, "emergent literature" is rather a concept used in the theory of literature.

Emergent Properties and Processes

An emergent behavior or emergent property can appear when a number of simple entities (agents) operate in an environment, forming more complex behaviors as a collective. If emergence happens over disparate size scales, then the reason is usually a causal relation across different scales. In other words, there is often a form of top-down feedback in systems with emergent properties. The processes from which emergent properties result may occur in either the observed or observing system, and can commonly be identified by their patterns of accumulating change, most generally called 'growth'. Emergent behaviours can occur because of intricate causal relations across different scales and feedback, known as interconnectivity. The emergent property itself may be either very predictable or unpredictable and unprecedented, and represent a new level of the system's evolution. The complex behaviour or properties are not a property of any single such entity, nor can they easily be predicted or deduced from behaviour in the lower-level entities, and might in fact be irreducible to such behavior. The shape and behaviour of a flock of birds or school of fish are good examples of emergent properties.

One reason why emergent behaviour is hard to predict is that the number of interactions between components of a system increases exponentially with the number of components, thus potentially allowing for many new and subtle types of behaviour to emerge. Emergence is often a product of particular patterns of interaction. Negative feedback introduces constraints that serve to fix structures or behaviours. In contrast, positive feedback promotes change, allowing local variations to grow into global patterns. Another way in which interactions leads to emergent properties is dual-phase evolution. This occurs where interactions are applied intermittently, leading to two phases: one in which patterns form or grow, the other in which they are refined or removed.

On the other hand, merely having a large number of interactions is not enough by itself to guarantee emergent behaviour; many of the interactions may be negligible or irrelevant, or may cancel each other out. In some cases, a large number of interactions can in fact work against the emergence of interesting behaviour, by creating a lot of "noise" to drown out any emerging "signal"; the emergent behaviour may need to be temporarily isolated from other interactions before it reaches enough critical mass to be self-supporting. Thus it is not just the sheer number of connections between components which encourages emergence; it is also how these connections are organised. A hierarchical organisation is one example that can generate emergent behaviour (a bureaucracy may behave in a way quite different from that of the individual humans in that bureaucracy); but perhaps more interestingly, emergent behaviour can also arise from more decentralized organisational structures, such as a marketplace. In some cases, the system has to reach a combined threshold of diversity, organisation, and connectivity before emergent behaviour appears.

Unintended consequences and side effects are closely related to emergent properties. Luc Steels writes: "A component has a particular functionality but this is not recognizable as a subfunction of the global functionality. Instead a component implements a behaviour whose side effect contributes to the global functionality [...] Each behaviour has a side effect and the sum of the side effects gives the desired functionality".(Steels 1990) In other words, the global or macroscopic functionality of a system with "emergent functionality" is the sum of all "side effects", of all emergent properties and functionalities.

Systems with emergent properties or emergent structures may appear to defy entropic principles and the second law of thermodynamics, because they form and increase order despite the lack of command and central control. This is possible because open systems can extract information and order out of the environment.

Emergence helps to explain why the fallacy of division is a fallacy.

Emergent Structures in Nature

Ripple patterns in a sand dune created by wind or water is an example of an emergent structure in nature

Giant's Causeway in Northern Ireland is an example of a complex emergent structure created by natural processes

Emergent structures can be found in many natural phenomena, from the physical to the biological domain. For example, the shape of weather phenomena such as hurricanes are emergent structures. The development and growth of complex, orderly crystals, as driven by the random motion of water molecules within a conducive natural environment, is another example of an emergent process, where randomness can give rise to complex and deeply attractive, orderly structures.

Water crystals forming on glass demonstrate an emergent, fractal natural process occurring under appropriate conditions of temperature and humidity.

However, crystalline structure and hurricanes are said to have a self-organizing phase.

It is useful to distinguish three forms of emergent structures. A *first-order* emergent structure occurs as a result of shape interactions (for example, hydrogen bonds in water molecules lead to surface tension). A *second-order* emergent structure involves shape interactions played out sequentially over time (for example, changing atmospheric conditions as a snowflake falls to the ground build upon and alter its form). Finally, a *third-order* emergent structure is a consequence of shape, time, and heritable instructions. For example, an organism's genetic code sets boundary conditions on the interaction of biological systems in space and time.

Non-living, Physical Systems

In physics, emergence is used to describe a property, law, or phenomenon which occurs at macroscopic scales (in space or time) but not at microscopic scales, despite the fact that a macroscopic system can be viewed as a very large ensemble of microscopic systems.

An emergent property need not be more complicated than the underlying non-emergent properties which generate it. For instance, the laws of thermodynamics are remarkably simple, even if the laws which govern the interactions between component particles are complex. The term emergence in physics is thus used not to signify complexity, but rather to distinguish which laws and concepts apply to macroscopic scales, and which ones apply to microscopic scales.

Some examples include:

- Classical mechanics: The laws of classical mechanics can be said to emerge as a limiting case from the rules of quantum mechanics applied to large enough masses. This is particularly strange since quantum mechanics is generally thought of as *more* complicated than classical mechanics.
- Friction: Forces between elementary particles are conservative. However, friction emerges when considering more complex structures of matter, whose surfaces can convert mechanical energy into heat energy when rubbed against each other. Similar considerations apply to other emergent concepts in continuum mechanics such as viscosity, elasticity, tensile strength, etc.
- Patterned ground: the distinct, and often symmetrical geometric shapes formed by ground material in periglacial regions.
- Statistical mechanics was initially derived using the concept of a large enough ensemble that fluctuations about the most likely distribution can be all but ignored. However, small clusters do not exhibit sharp first order phase transitions such as melting, and at the boundary it is not possible to completely categorize the cluster as a liquid or solid, since these concepts are (without extra definitions) only applicable to macroscopic systems. Describing a system using statistical mechanics methods is much simpler than using a low-level atomistic approach.
- Electrical networks: The bulk conductive response of binary (RC) electrical networks with random arrangements can be seen as emergent properties of such physical systems. Such arrangements can be used as simple physical prototypes for deriving mathematical formulae for the emergent responses of complex systems.
- Weather

Temperature is sometimes used as an example of an emergent macroscopic behaviour. In classical dynamics, a *snapshot* of the instantaneous momenta of a large number of particles at equilibrium is sufficient to find the average kinetic energy per degree of freedom which is proportional to the temperature. For a small number of particles the instantaneous momenta at a given time are not statistically sufficient to determine the temperature of the system. However, using the ergodic hypothesis, the temperature can still be obtained to arbitrary precision by further averaging the momenta over a long enough time.

Convection in a liquid or gas is another example of emergent macroscopic behaviour that makes sense only when considering differentials of temperature. Convection cells, particularly Bénard cells, are an example of a self-organizing system (more specifically, a dissipative system) whose structure is determined both by the constraints of the system and by random perturbations: the possible realizations of the shape and size of the cells depends on the temperature gradient as well as the nature of the fluid and shape of the container, but which configurations are actually realized is due to random perturbations (thus these systems exhibit a form of symmetry breaking).

In some theories of particle physics, even such basic structures as mass, space, and time are viewed as emergent phenomena, arising from more fundamental concepts such as the Higgs boson or strings. In some interpretations of quantum mechanics, the perception of a deterministic reality, in which all objects have a definite position, momentum, and so forth, is actually an emergent phenomenon, with the true state of matter being described instead by a wavefunction which need not have a single position or momentum. Most of the laws of physics themselves as we experience them today appear to have emerged during the course of time making emergence the most fundamental principle in the universe and raising the question of what might be the most fundamental law of physics from which all others emerged. Chemistry can in turn be viewed as an emergent property of the laws of physics. Biology (including biological evolution) can be viewed as an emergent property of the laws of chemistry. Similarly, psychology could be understood as an emergent property of neurobiological laws. Finally, free-market theories understand economy as an emergent feature of psychology.

According to Laughlin (2005), for many particle systems, nothing can be calculated exactly from the microscopic equations, and macroscopic systems are characterised by broken symmetry: the symmetry present in the microscopic equations is not present in the macroscopic system, due to phase transitions. As a result, these macroscopic systems are described in their own terminology, and have properties that do not depend on many microscopic details. This does not mean that the microscopic interactions are irrelevant, but simply that you do not see them anymore — you only see a renormalized effect of them. Laughlin is a pragmatic theoretical physicist: if you cannot, possibly ever, calculate the broken symmetry macroscopic properties from the microscopic equations, then what is the point of talking about reducibility?

Living, Biological Systems

Emergence and Evolution

Life is a major source of complexity, and evolution is the major process behind the varying forms of life. In this view, evolution is the process describing the growth of complexity in the natural world and in speaking of the emergence of complex living beings and life-forms, this view refers therefore to processes of sudden changes in evolution.

Life is thought to have emerged in the early RNA world when RNA chains began to express the basic conditions necessary for natural selection to operate as conceived by Darwin: heritability, variation of type, and competition for limited resources. Fitness of an RNA replicator (its per capita rate of increase) would likely be a function of adaptive capacities that were intrinsic (in the sense that they were determined by the nucleotide sequence) and the availability of resources. The three primary adaptive capacities may have been (1) the capacity to replicate with moderate fidelity (giving rise to both heritability and variation of type); (2) the capacity to avoid decay; and (3) the capacity to acquire and process resources. These capacities would have been determined initially by the folded configurations of the RNA replicators that, in turn, would be encoded in their individual nucleotide sequences. Competitive success among different replicators would have depended on the relative values of these adaptive capacities.

Regarding causality in evolution Peter Corning observes:

Synergistic effects of various kinds have played a major causal role in the evolutionary process generally and in the evolution of cooperation and complexity in particular... Natural selection is often portrayed as a "mechanism", or is personified as a causal agency... In reality, the differential "selection" of a trait, or an adaptation, is a consequence of the functional effects it produces in relation to the survival and reproductive success of a given organism in a given environment. It is these functional effects that are ultimately responsible for the trans-generational continuities and changes in nature.(Corning 2002).

Per his definition of emergence, Corning also addresses emergence and evolution:

[In] evolutionary processes, causation is iterative; effects are also causes. And this is equally true of the synergistic effects produced by emergent systems. In other words, emergence itself... has been the underlying cause of the evolution of emergent phenomena in biological evolution; it is the synergies produced by organized systems that are the key.(Corning 2002).

Swarming is a well-known behaviour in many animal species from marching locusts to schooling fish to flocking birds. Emergent structures are a common strategy found in many animal groups: colonies of ants, mounds built by termites, swarms of bees, shoals/schools of fish, flocks of birds, and herds/packs of mammals.

An example to consider in detail is an ant colony. The queen does not give direct orders and does not tell the ants what to do. Instead, each ant reacts to stimuli in the form of chemical scent from larvae, other ants, intruders, food and buildup of waste, and leaves behind a chemical trail, which, in turn, provides a stimulus to other ants. Here each ant is an autonomous unit that reacts depending only on its local environment and the genetically encoded rules for its variety of ant. Despite the lack of centralized decision making, ant colonies exhibit complex behavior and have even demonstrated the ability to solve geometric problems. For example, colonies routinely find the maximum distance from all colony entrances to dispose of dead bodies.

It appears that environmental factors may play a role in influencing emergence. Research suggests induced emergence of the bee species Macrotera portalis. In this species, the bees emerge in a pattern consistent with rainfall. Specifically, the pattern of emergence is consistent with southwestern deserts' late summer rains and lack of activity in the spring.

Organization of Life

A broader example of emergent properties in biology is viewed in the biological organisation of life, ranging from the subatomic level to the entire biosphere. For example, individual atoms can be combined to form molecules such as polypeptide chains, which in turn fold and refold to form proteins, which in turn create even more complex structures. These proteins, assuming their functional status from their spatial conformation, interact together and with other molecules to achieve higher biological functions and eventually create an organism. Another example is how cascade phenotype reactions, as detailed in chaos theory, arise from individual genes mutating respective positioning. At the highest level, all the biological communities in the world form the biosphere, where its human participants form societies, and the complex interactions of meta-social systems such as the stock market.

Emergence of Mind

Among the considered phenomena in the evolutionary account of life, as a continuous history, marked by stages at which fundamentally new forms have appeared - the origin of sapiens intelligence. The emergence of mind and its evolution is researched and considered as a separate phenomenon in a special system knowledge noogenesis.

In Humanity

Spontaneous Order

Groups of human beings, left free to each regulate themselves, tend to produce spontaneous order, rather than the meaningless chaos often feared. This has been observed in society at least since Chuang Tzu in ancient China. A classic traffic roundabout is a good example, with cars moving in and out with such effective organization that some modern cities have begun replacing stoplights at problem intersections with traffic circles, and getting better results. Open-source software and Wiki projects form an even more compelling illustration.

Emergent processes or behaviours can be seen in many other places, such as cities, cabal and market-dominant minority phenomena in economics, organizational phenomena in computer simulations and cellular automata. Whenever you have a multitude of individuals interacting with one another, there often comes a moment when disorder gives way to order and something new emerges: a pattern, a decision, a structure, or a change in direction (Miller 2010, 29).

Economics

The stock market (or any market for that matter) is an example of emergence on a grand scale. As a whole it precisely regulates the relative security prices of companies across the world, yet it has no leader; when no central planning is in place, there is no one entity which controls the workings of the entire market. Agents, or investors, have knowledge of only a limited number of companies within their portfolio, and must follow the regulatory rules of the market and analyse the transactions individually or in large groupings. Trends and patterns emerge which are studied intensively by technical analysts.. Emergence has been shown to occur in certain econophysics models of economics systems.

World Wide Web and the Internet

The World Wide Web is a popular example of a decentralized system exhibiting emergent properties. There is no central organization rationing the number of links, yet the number of links pointing to each page follows a power law in which a few pages are linked to many times and most pages are seldom linked to. A related property of the network of links in the World Wide Web is that almost any pair of pages can be connected to each other through a relatively short chain of links. Although relatively well known now, this property was initially unexpected in an unregulated network. It is shared with many other types of networks called small-world networks.(Barabasi, Jeong, & Albert 1999, pp. 130–31).

Internet traffic can also exhibit some seemingly emergent properties. In the congestion control mechanism, TCP flows can become globally synchronized at bottlenecks, simultaneously increasing and then decreasing throughput in coordination. Congestion, widely regarded as a nuisance, is possibly an emergent property of the spreading of bottlenecks across a network in high traffic flows which can be considered as a phase transition.

Another important example of emergence in web-based systems is social bookmarking (also called collaborative tagging). In social bookmarking systems, users assign tags to resources shared with other users, which gives rise to a type of information organisation that emerges from this crowdsourcing process. Recent research which analyzes empirically the complex dynamics of such systems has shown that consensus on stable distributions and a simple form of shared vocabularies does indeed emerge, even in the absence of a central controlled vocabulary. Some believe that this could be because users who contribute tags all use the same language, and they share similar semantic structures underlying the choice of words. The convergence in social tags may therefore be interpreted as the emergence of structures as people who have similar semantic interpretation collaboratively index online information, a process called semantic imitation.

Architecture and Cities

Traffic patterns in cities can be seen as an example of spontaneous order

Emergent structures appear at many different levels of organization or as spontaneous order. Emergent self-organization appears frequently in cities where no planning or zoning entity predetermines the layout of the city.(Krugman 1996, pp. 9–29) The interdisciplinary study of emergent behaviors is not generally considered a homogeneous field, but divided across its application or problem domains.

Architects and Landscape Architects may not design all the pathways of a complex of buildings. Instead they might let usage patterns emerge and then place pavement where pathways have become worn, such as a desire path.

The on-course action and vehicle progression of the 2007 Urban Challenge could possibly be regarded as an example of cybernetic emergence. Patterns of road use, indeterministic obstacle clearance times, etc. will work together to form a complex emergent pattern that can not be deterministically planned in advance.

The architectural school of Christopher Alexander takes a deeper approach to emergence attempting to rewrite the process of urban growth itself in order to affect form, establishing a new methodology of planning and design tied to traditional practices, an Emergent Urbanism. Urban emergence has also been linked to theories of urban complexity (Batty 2005) and urban evolution. (Marshall 2009).

Building ecology is a conceptual framework for understanding architecture and the built environment as the interface between the dynamically interdependent elements of buildings, their occupants, and the larger environment. Rather than viewing buildings as inanimate or static objects, building ecologist Hal Levin views them as interfaces or intersecting domains of living and non-living systems. The microbial ecology of the indoor environment is strongly dependent on the building materials, occupants, contents, environmental context and the indoor and outdoor climate. The strong relationship between atmospheric chemistry and indoor air quality and the chemical reactions occurring indoors. The chemicals may be nutrients, neutral or biocides for the microbial organisms. The microbes produce chemicals that affect the building materials and occupant health and well being. Humans manipulate the ventilation, temperature and humidity to achieve comfort with the concomitant effects on the microbes that populate and evolve.

Eric Bonabeau's attempt to define emergent phenomena is through traffic: "traffic jams are actually very complicated and mysterious. On an individual level, each driver is trying to get somewhere and is following (or breaking) certain rules, some legal (the speed limit) and others societal or personal (slow down to let another driver change into your lane). But a traffic jam is a separate and distinct entity that emerges from those individual behaviors. Gridlock on a highway, for example, can travel backward for no apparent reason, even as the cars are moving forward." He has also likened emergent phenomena to the analysis of market trends and employee behavior.

Computational emergent phenomena have also been utilized in architectural design processes, for example for formal explorations and experiments in digital materiality.

Computer AI

Some artificially intelligent (AI) computer applications utilize emergent behavior for animation. One example is Boids, which mimics the swarming behavior of birds.

Language

It has been argued that the structure and regularity of language grammar, or at least language change, is an emergent phenomenon (Hopper 1998). While each speaker merely tries to reach his or her own communicative goals, he or she uses language in a particular way. If enough speakers behave in that way, language is changed (Keller 1994). In a wider sense, the norms of a language, i.e. the linguistic conventions of its speech society, can be seen as a system emerging from long-time participation in communicative problem-solving in various social circumstances (Määttä 2000).

Emergent Change Processes

Within the field of group facilitation and organization development, there have been a number of new group processes that are designed to maximize emergence and self-organization, by offering a minimal set of effective initial conditions. Examples of these processes include SEED-SCALE,

Appreciative Inquiry, Future Search, the World Cafe or Knowledge Cafe, Open Space Technology, and others (Holman, 2010).

Electronic Band Structure

In solid-state physics, the electronic band structure (or simply band structure) of a solid describes the range of energies that an electron within the solid may have (called *energy bands, allowed bands*, or simply *bands*) and ranges of energy that it may not have (called *band gaps* or *forbidden bands*).

Band theory derives these bands and band gaps by examining the allowed quantum mechanical wave functions for an electron in a large, periodic lattice of atoms or molecules. Band theory has been successfully used to explain many physical properties of solids, such as electrical resistivity and optical absorption, and forms the foundation of the understanding of all solid-state devices (transistors, solar cells, etc.).

Why Bands and Band Gaps Occur

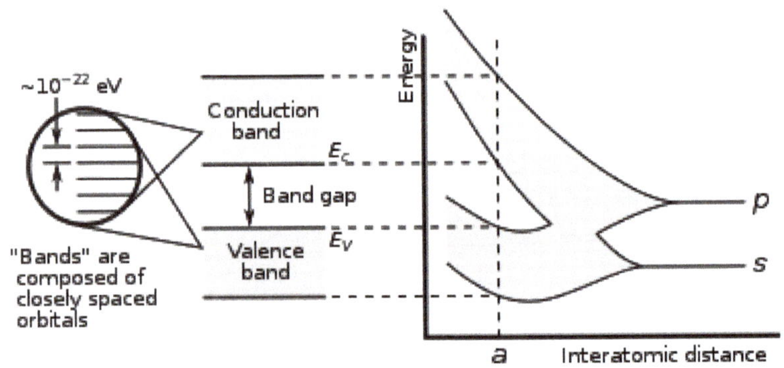

Showing how electronic band structure comes about by the hypothetical example of a large number of carbon atoms being brought together to form a diamond crystal.

The graph *(right)* shows the energy levels as a function of the spacing between atoms. When the atoms are far apart *(right side of graph)* each atom has valence atomic orbitals p and s which have the same energy. However when the atoms come closer together their orbitals begin to overlap. Due to the Pauli Exclusion Principle each atomic orbital splits into N molecular orbitals each with a different energy, where N is the number of atoms in the crystal. Since N is such a large number, adjacent orbitals are extremely close together in energy so the orbitals can be considered a continuous energy band. a is the atomic spacing in an actual crystal of diamond. At that spacing the orbitals form two bands, called the valence and conduction bands, with a 5.5 eV band gap between them.

The electrons of a single, isolated atom occupy atomic orbitals each of which has a discrete energy level. When two atoms join together to form into a molecule, their atomic orbitals overlap. The Pauli exclusion principle dictates that no two electrons can have the same quantum numbers in a molecule. So if two identical atoms combine to form a diatomic molecule, each atomic orbital splits into two molecular orbitals of different energy, allowing the electrons in the former atomic orbitals to occupy the new orbital structure without any having the same energy.

Similarly if a large number N of identical atoms come together to form a solid, such as a crystal lattice, the atoms' atomic orbitals overlap. Since the Pauli exclusion principle dictates that no two electrons in the solid have the same quantum numbers, each atomic orbital splits into N discrete molecular orbitals, each with a different energy. Since the number of atoms in a macroscopic piece of solid is a very large number ($N \sim 10^{22}$) the number of orbitals is very large and thus they are very closely spaced in energy (of the order of 10^{-22} eV). The energy of adjacent levels is so close together that they can be considered as a continuum, an energy band.

This formation of bands is mostly a feature of the outermost electrons (valence electrons) in the atom, which are the ones responsible for chemical bonding and electrical conductivity. The inner electron orbitals do not overlap to a significant degree, so their bands are very narrow.

Band gaps are essentially leftover ranges of energy not covered by any band, a result of the finite widths of the energy bands. The bands have different widths, with the widths depending upon the degree of overlap in the atomic orbitals from which they arise. Two adjacent bands may simply not be wide enough to fully cover the range of energy. For example, the bands associated with core orbitals (such as 1s electrons) are extremely narrow due to the small overlap between adjacent atoms. As a result, there tend to be large band gaps between the core bands. Higher bands involve comparatively larger orbitals with more overlap, becoming progressively wider at higher energies so that there are no band gaps at higher energies.

Basic concepts

Assumptions and Limits of Band Structure Theory

Band theory is only an approximation to the quantum state of a solid, which applies to solids consisting of many identical atoms or molecules bonded together. These are the assumptions necessary for band theory to be valid:

- *Infinite-size system*: For the bands to be continuous, the piece of material must consist of a large number of atoms. Since a macroscopic piece of material contains on the order of 10^{22} atoms, this is not a serious restriction; band theory even applies to microscopic-sized transistors in integrated circuits. With modifications, the concept of band structure can also be extended to systems which are only "large" along some dimensions, such as two-dimensional electron systems.
- *Homogeneous system*: Band structure is an intrinsic property of a material, which assumes that the material is homogeneous. Practically, this means that the chemical makeup of the material must be uniform throughout the piece.
- *Non-interactivity*: The band structure describes "single electron states". The existence of these states assumes that the electrons travel in a static potential without dynamically interacting with lattice vibrations, other electrons, photons, etc.

The above assumptions are broken in a number of important practical situations, and the use of band structure requires one to keep a close check on the limitations of band theory:

- Inhomogeneities and interfaces: Near surfaces, junctions, and other inhomogeneities, the bulk band structure is disrupted. Not only are there local small-scale disruptions (e.g., surface states or dopant states inside the band gap), but also local charge imbalances. These charge imbalances have electrostatic effects that extend deeply into semiconductors, insulators, and the vacuum.

- Along the same lines, most electronic effects (capacitance, electrical conductance, electric-field screening) involve the physics of electrons passing through surfaces and/or near interfaces. The full description of these effects, in a band structure picture, requires at least a rudimentary model of electron-electron interactions.
- Small systems: For systems which are small along every dimension (e.g., a small molecule or a quantum dot), there is no continuous band structure. The crossover between small and large dimensions is the realm of mesoscopic physics.
- Strongly correlated materials (for example, Mott insulators) simply cannot be understood in terms of single-electron states. The electronic band structures of these materials are poorly defined (or at least, not uniquely defined) and may not provide useful information about their physical state.

Crystalline Symmetry and Wavevectors

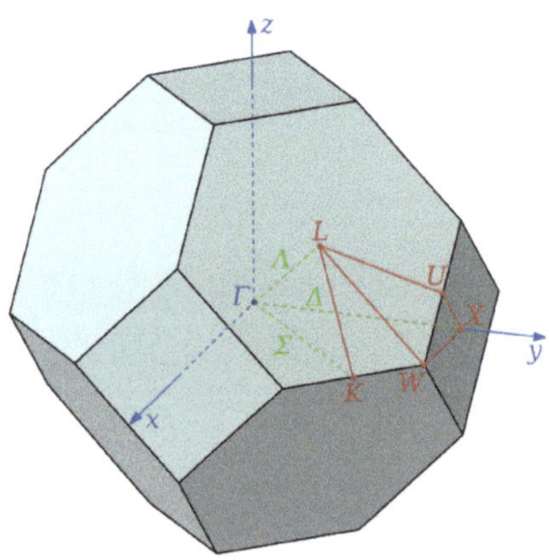

Fig a. Brillouin zone of a face-centered cubic lattice showing labels for special symmetry points

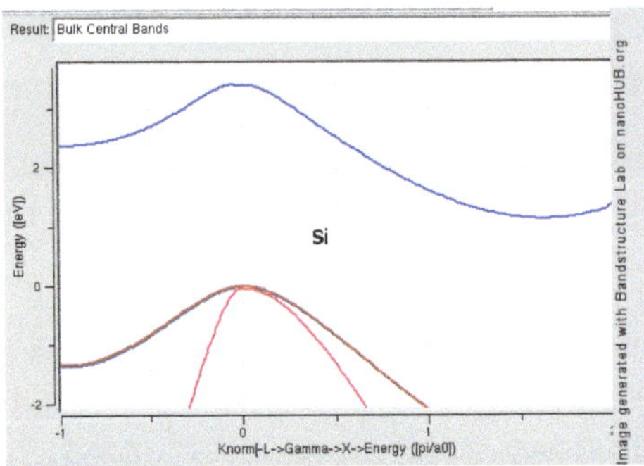

Fig b. Band structure plot for Si, Ge, GaAs and InAs generated with tight binding model. Note that Si and Ge are indirect band gap materials, while GaAs and InAs are direct.

Band structure calculations take advantage of the periodic nature of a crystal lattice, exploiting its symmetry. The single-electron Schrödinger equation is solved for an electron in a lattice-periodic potential, giving Bloch waves as solutions:

$$\psi_{n\mathbf{k}}(\mathbf{r}) = e^{i\mathbf{k}\cdot\mathbf{r}} u_{n\mathbf{k}}(\mathbf{r}),$$

where k is called the wavevector. For each value of k, there are multiple solutions to the Schrödinger equation labelled by n, the band index, which simply numbers the energy bands. Each of these energy levels evolves smoothly with changes in k, forming a smooth band of states. For each band we can define a function $E_n(\mathrm{k})$, which is the dispersion relation for electrons in that band.

The wavevector takes on any value inside the Brillouin zone, which is a polyhedron in wavevector space that is related to the crystal's lattice. Wavevectors outside the Brillouin zone simply correspond to states that are physically identical to those states within the Brillouin zone. Special high symmetry points/lines in the Brillouin zone are assigned labels like Γ, Δ, Λ, Σ (see Fig a).

It is difficult to visualize the shape of a band as a function of wavevector, as it would require a plot in four-dimensional space, E vs. k_x, k_y, k_z. In scientific literature it is common to see band structure plots which show the values of $E_n(\mathrm{k})$ for values of k along straight lines connecting symmetry points, often labelled Δ, Λ, Σ, or , , and , respectively. Another method for visualizing band structure is to plot a constant-energy isosurface in wavevector space, showing all of the states with energy equal to a particular value. The isosurface of states with energy equal to the Fermi level is known as the Fermi surface.

Energy band gaps can be classified using the wavevectors of the states surrounding the band gap:

- Direct band gap: the lowest-energy state above the band gap has the same k as the highest-energy state beneath the band gap.
- Indirect band gap: the closest states above and beneath the band gap do not have the same k value.

Asymmetry: Band Structures in Non-crystalline Solids

Although electronic band structures are usually associated with crystalline materials, quasi-crystalline and amorphous solids may also exhibit band structures. These are somewhat more difficult to study theoretically since they lack the simple symmetry of a crystal, and it is not usually possible to determine a precise dispersion relation. As a result, virtually all of the existing theoretical work on the electronic band structure of solids has focused on crystalline materials.

Density of States

The density of states function $g(E)$ is defined as the number of electronic states per unit volume, per unit energy, for electron energies near E.

The density of states function is important for calculations of effects based on band theory. In Fermi's Golden Rule, a calculation for the rate of optical absorption, it provides both the number of excitable electrons and the number of final states for an electron. It appears in calculations of electrical conductivity where it provides the number of mobile states, and in computing electron scattering rates where it provides the number of final states after scattering.

For energies inside a band gap, $g(E) = 0$.

Filling of Bands

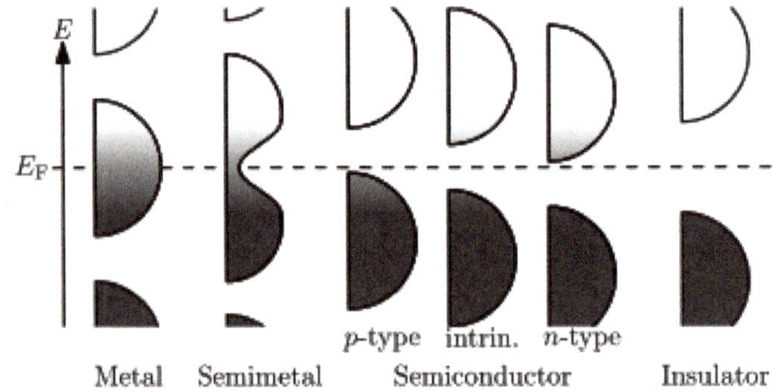

Filling of the electronic states in various types of materials at equilibrium. Here, height is energy while width is the density of available states for a certain energy in the material listed. The shade follows the Fermi–Dirac distribution (black = all states filled, white = no state filled). In metals and semimetals the Fermi level E_F lies inside at least one band. In insulators and semiconductors the Fermi level is inside a band gap; however, in semiconductors the bands are near enough to the Fermi level to be thermally populated with electrons or holes.

At thermodynamic equilibrium, the likelihood of a state of energy *E* being filled with an electron is given by the Fermi–Dirac distribution, a thermodynamic distribution that takes into account the Pauli exclusion principle:

$$f(E) = \frac{1}{1 + e^{(E-\mu)/k_B T}}$$

where:

- $k_B T$ is the product of Boltzmann's constant and temperature, and
- μ is the total chemical potential of electrons, or *Fermi level* (in semiconductor physics, this quantity is more often denoted E_F). The Fermi level of a solid is directly related to the voltage on that solid, as measured with a voltmeter. Conventionally, in band structure plots the Fermi level is taken to be the zero of energy (an arbitrary choice).

The density of electrons in the material is simply the integral of the Fermi–Dirac distribution times the density of states:

$$N / V = \int_{-\infty}^{\infty} g(E) f(E) dE$$

Although there are an infinite number of bands and thus an infinite number of states, there are only a finite number of electrons to place in these bands. The preferred value for the number of electrons is a consequence of electrostatics: even though the surface of a material can be charged, the internal bulk of a material prefers to be charge neutral. The condition of charge neutrality means that *N/V* must match the density of protons in the material. For this to occur, the material electrostatically adjusts itself, shifting its band structure up or down in energy (thereby shifting *g*(*E*)), until it is at the correct equilibrium with respect to the Fermi level.

Names of Bands Near the Fermi Level (Conduction Band, Valence Band)

A solid has an infinite number of allowed bands, just as an atom has infinitely many energy levels. However, most of the bands simply have too high energy, and are usually disregarded under ordinary circumstances. Conversely, there are very low energy bands associated with the core orbitals (such as 1s electrons). These low-energy *core bands* are also usually disregarded since they remain filled with electrons at all times, and are therefore inert. Likewise, materials have several band gaps throughout their band structure.

The most important bands and band gaps—those relevant for electronics and optoelectronics—are those with energies near the Fermi level. The bands and band gaps near the Fermi level are given special names, depending on the material:

- In a semiconductor or band insulator, the Fermi level is surrounded by a band gap, referred to as *the* band gap (to distinguish it from the other band gaps in the band structure). The closest band above the band gap is called *the conduction band*, and the closest band beneath the band gap is called *the valence band*. The name "valence band" was coined by analogy to chemistry, since in many semiconductors the valence band is built out of the valence orbitals.
- In a metal or semimetal, the Fermi level is inside of one or more allowed bands. In semimetals the bands are usually referred to as "conduction band" or "valence band" depending on whether the charge transport is more electron-like or hole-like, by analogy to semiconductors. In many metals, however, the bands are neither electron-like nor hole-like, and often just called "valence band" as they are made of valence orbitals. The band gaps in a metal's band structure are not important for low energy physics, since they are too far from the Fermi level.

Theory of Band Structures in Crystals

The ansatz is the special case of electron waves in a periodic crystal lattice using Bloch waves as treated generally in the dynamical theory of diffraction. Every crystal is a periodic structure which can be characterized by a Bravais lattice, and for each Bravais lattice we can determine the reciprocal lattice, which encapsulates the periodicity in a set of three reciprocal lattice vectors (b_1,b_2,b_3). Now, any periodic potential V(**r**) which shares the same periodicity as the direct lattice can be expanded out as a Fourier series whose only non-vanishing components are those associated with the reciprocal lattice vectors. So the expansion can be written as:

$$V(\mathbf{r}) = \sum_{\mathbf{K}} V_{\mathbf{K}} e^{i\mathbf{K}\cdot\mathbf{r}}$$

where $K = m_1b_1 + m_2b_2 + m_3b_3$ for any set of integers (m_1,m_2,m_3).

From this theory, an attempt can be made to predict the band structure of a particular material, however most ab initio methods for electronic structure calculations fail to predict the observed band gap.

Nearly Free Electron Approximation

In the nearly free electron approximation, interactions between electrons are completely ignored. This approximation allows use of Bloch's Theorem which states that electrons in a periodic potential have wavefunctions and energies which are periodic in wavevector up to a constant phase shift between neighboring reciprocal lattice vectors. The consequences of periodicity are described mathematically by the Bloch wavefunction:

$$\Psi_{n,\mathbf{k}}(\mathbf{r}) = e^{i\mathbf{k}\cdot\mathbf{r}} u_n(\mathbf{r})$$

where the function $u_n(\mathbf{r})$ is periodic over the crystal lattice, that is,

$$u_n(\mathbf{r}) = u_n(\mathbf{r} - \mathbf{R}).$$

Here index n refers to the *n-th* energy band, wavevector k is related to the direction of motion of the electron, r is the position in the crystal, and R is the location of an atomic site.

The NFE model works particularly well in materials like metals where distances between neighbouring atoms are small. In such materials the overlap of atomic orbitals and potentials on neighbouring atoms is relatively large. In that case the wave function of the electron can be approximated by a (modified) plane wave. The band structure of a metal like aluminium even gets close to the empty lattice approximation.

Tight Binding Model

The opposite extreme to the nearly free electron approximation assumes the electrons in the crystal behave much like an assembly of constituent atoms. This tight binding model assumes the solution to the time-independent single electron Schrödinger equation Ψ is well approximated by a linear combination of atomic orbitals $\psi_n(\mathbf{r})$.

$$\Psi(\mathbf{r}) = \sum_{n,\mathbf{R}} b_{n,\mathbf{R}} \psi_n(\mathbf{r} - \mathbf{R}),$$

where the coefficients $b_{n,\mathbf{R}}$ are selected to give the best approximate solution of this form. Index n refers to an atomic energy level and R refers to an atomic site. A more accurate approach using this idea employs Wannier functions, defined by:

$$a_n(\mathbf{r} - \mathbf{R}) = \frac{V_C}{(2\pi)^3} \int_{BZ} d\mathbf{k} e^{-i\mathbf{k}\cdot(\mathbf{R}-\mathbf{r})} u_{n\mathbf{k}} ;$$

in which $u_{n\mathbf{k}}$ is the periodic part of the Bloch wave and the integral is over the Brillouin zone. Here index n refers to the n-th energy band in the crystal. The Wannier functions are localized near atomic sites, like atomic orbitals, but being defined in terms of Bloch functions they are accurately related to solutions based upon the crystal potential. Wannier functions on different atomic sites R are orthogonal. The Wannier functions can be used to form the Schrödinger solution for the n-th energy band as:

$$\Psi_{n,\mathbf{k}}(\mathbf{r}) = \sum_{\mathbf{R}} e^{-i\mathbf{k}\cdot(\mathbf{R}-\mathbf{r})} a_n(\mathbf{r} - \mathbf{R}).$$

The TB model works well in materials with limited overlap between atomic orbitals and potentials on neighbouring atoms. Band structures of materials like Si, GaAs, SiO_2 and diamond for instance are well described by TB-Hamiltonians on the basis of atomic sp^3 orbitals. In transition metals a mixed TB-NFE model is used to describe the broad NFE conduction band and the narrow embedded TB d-bands. The radial functions of the atomic orbital part of the Wannier functions are most easily calculated by the use of pseudopotential methods. NFE, TB or combined NFE-TB band structure calculations, sometimes extended with wave function approximations based on pseudo-potential methods, are often used as an economic starting point for further calculations.

KKR Model

The simplest form of this approximation centers non-overlapping spheres (referred to as *muffin tins*) on the atomic positions. Within these regions, the potential experienced by an electron is approximated to be spherically symmetric about the given nucleus. In the remaining interstitial region, the screened potential is approximated as a constant. Continuity of the potential between the atom-centered spheres and interstitial region is enforced.

A variational implementation was suggested by Korringa and by Kohn and Rostocker, and is often referred to as the *KKR model.*

Density-functional Theory

In recent physics literature, a large majority of the electronic structures and band plots are calculated using density-functional theory (DFT), which is not a model but rather a theory, i.e., a microscopic first-principles theory of condensed matter physics that tries to cope with the electron-electron many-body problem via the introduction of an exchange-correlation term in the functional of the electronic density. DFT-calculated bands are in many cases found to be in agreement with experimentally measured bands, for example by angle-resolved photoemission spectroscopy (ARPES). In particular, the band shape is typically well reproduced by DFT. But there are also systematic errors in DFT bands when compared to experiment results. In particular, DFT seems to systematically underestimate by about 30-40% the band gap in insulators and semiconductors.

It is commonly believed that DFT is a theory to predict ground state properties of a system only (e.g. the total energy, the atomic structure, etc.), and that excited state properties cannot be determined by DFT. This is a misconception. In principle, DFT can determine any property (ground state or excited state) of a system given a functional that maps the ground state density to that property. This is the essence of the Hohenberg–Kohn theorem. In practice, however, no known functional exists that maps the ground state density to excitation energies of electrons within a material. Thus, what in the literature is quoted as a DFT band plot is a representation of the DFT Kohn–Sham energies, i.e., the energies of a fictive non-interacting system, the Kohn–Sham system, which has no physical interpretation at all. The Kohn–Sham electronic structure must not be confused with the real, quasiparticle electronic structure of a system, and there is no Koopman's theorem holding for Kohn–Sham energies, as there is for Hartree–Fock energies, which can be truly considered as an approximation for quasiparticle energies. Hence, in principle, Kohn–Sham based DFT is not a band theory, i.e., not a theory suitable for calculating bands and band-plots. In principle time-dependent DFT can be used to calculate the true band structure although in practice this is often difficult. A popular approach is the use of hybrid functionals, which incorporate a portion of Hartree–Fock exact exchange; this produces a substantial improvement in predicted bandgaps of semiconductors, but is less reliable for metals and wide-bandgap materials.

Green's Function Methods and the Ab Initio GW Approximation

To calculate the bands including electron-electron interaction many-body effects, one can resort to so-called Green's function methods. Indeed, knowledge of the Green's function of a system provides both ground (the total energy) and also excited state observables of the system. The poles of the Green's function are the quasiparticle energies, the bands of a solid. The Green's function can be calculated by solving the Dyson equation once the self-energy of the system is known. For real systems like solids, the self-energy is a very complex quantity and usually approximations are needed to solve the problem. One such approximation is the GW approximation, so called from the mathematical form the self-energy takes as the product $\Sigma = GW$ of the Green's function G and the dynamically screened interaction W. This approach is more pertinent when addressing the calculation of band plots (and also quantities beyond, such as the spectral function) and can also be formulated in a completely *ab initio* way. The GW approximation seems to provide band gaps of insulators and semiconductors in agreement with experiment, and hence to correct the systematic DFT underestimation.

Mott Insulators

Although the nearly free electron approximation is able to describe many properties of electron band structures, one consequence of this theory is that it predicts the same number of electrons in each unit cell. If the number of electrons is odd, we would then expect that there is an unpaired electron in each unit cell, and thus that the valence band is not fully occupied, making the material a conductor. However, materials such as CoO that have an odd number of electrons per unit cell are insulators, in direct conflict with this result. This kind of material is known as a Mott insulator, and requires inclusion of detailed electron-electron interactions (treated only as an averaged effect on the crystal potential in band theory) to explain the discrepancy. The Hubbard model is an approximate theory that can include these interactions. It can be treated non-perturbatively within the so-called dynamical mean field theory, which attempts to bridge the gap between the nearly free electron approximation and the atomic limit. Formally, however, the states are not non-interacting in this case and the concept of a band structure is not adequate to describe these cases.

Others

Calculating band structures is an important topic in theoretical solid state physics. In addition to the models mentioned above, other models include the following:

- Empty lattice approximation: the "band structure" of a region of free space that has been divided into a lattice.
- k·p perturbation theory is a technique that allows a band structure to be approximately described in terms of just a few parameters. The technique is commonly used for semiconductors, and the parameters in the model are often determined by experiment.
- The Kronig-Penney Model, a one-dimensional rectangular well model useful for illustration of band formation. While simple, it predicts many important phenomena, but is not quantitative.
- Hubbard model.

The band structure has been generalised to wavevectors that are complex numbers, resulting in what is called a *complex band structure*, which is of interest at surfaces and interfaces.

Each model describes some types of solids very well, and others poorly. The nearly free electron model works well for metals, but poorly for non-metals. The tight binding model is extremely accurate for ionic insulators, such as metal halide salts (e.g. NaCl).

Band Diagrams

To understand how band structure changes relative to the Fermi level in real space, a band structure plot is often first simplified in the form of a band diagram. In a band diagram the vertical axis is energy while the horizontal axis represents real space. Horizontal lines represent energy levels, while blocks represent energy bands. When the horizontal lines in these diagram are slanted then the energy of the level or band changes with distance. Diagrammatically, this depicts the presence of an electric field within the crystal system. Band diagrams are useful in relating the general band structure properties of different materials to one another when placed in contact with each other.

Symmetry Breaking

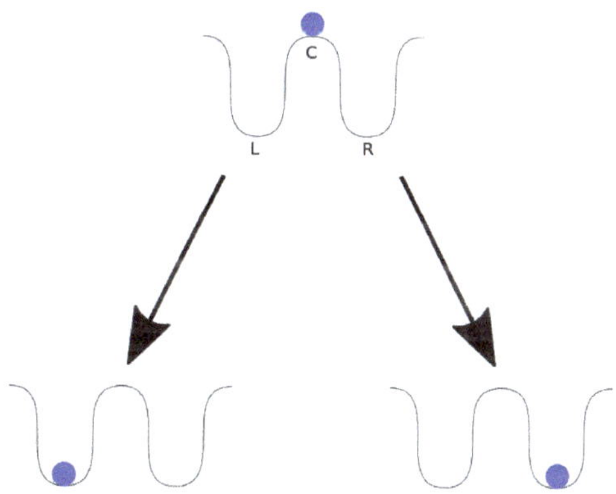

A ball is initially located at the top of the central hill (C). This position is an unstable equilibrium: a very small perturbation will cause it to fall to one of the two stable wells left (L) or (R). Even if the hill is symmetric and there is no reason for the ball to fall on either side, the observed final state is not symmetric.

In physics, symmetry breaking is a phenomenon in which (infinitesimally) small fluctuations acting on a system crossing a critical point decide the system's fate, by determining which branch of a bifurcation is taken. To an outside observer unaware of the fluctuations (or "noise"), the choice will appear arbitrary. This process is called symmetry "breaking", because such transitions usually bring the system from a symmetric but disorderly state into one or more definite states. Symmetry breaking is thought to play a major role in pattern formation.

In 1972, Nobel laureate P.W. Anderson used the idea of symmetry breaking to show some of the drawbacks of reductionism in his paper titled "More is different" in *Science*.

Symmetry breaking can be distinguished into two types, explicit symmetry breaking and spontaneous symmetry breaking, characterized by whether the equations of motion fail to be invariant or the ground state fails to be invariant.

Explicit Symmetry Breaking

In explicit symmetry breaking, the equations of motion describing a system are variant under the broken symmetry.

Spontaneous Symmetry Breaking

In spontaneous symmetry breaking, the equations of motion of the system are invariant, but the system is not because the background (spacetime) of the system, its vacuum, is non-invariant. Such a symmetry breaking is parametrized by an order parameter. A special case of this type of symmetry breaking is dynamical symmetry breaking.

Examples

Symmetry breaking can cover any of the following scenarios:

- The breaking of an exact symmetry of the underlying laws of physics by the random formation of some structure;
- A situation in physics in which a minimal energy state has less symmetry than the system itself;
- Situations where the actual state of the system does not reflect the underlying symmetries of the dynamics because the manifestly symmetric state is unstable (stability is gained at the cost of local asymmetry);
- Situations where the equations of a theory may have certain symmetries, though their solutions may not (the symmetries are "hidden").

One of the first cases of broken symmetry discussed in the physics literature is related to the form taken by a uniformly rotating body of incompressible fluid in gravitational and hydrostatic equilibrium. Jacobi and soon later Liouville, in 1834, discussed the fact that a tri-axial ellipsoid was an equilibrium solution for this problem when the kinetic energy compared to the gravitational energy of the rotating body exceeded a certain critical value. The axial symmetry presented by the McLaurin spheroids is broken at this bifurcation point. Furthermore, above this bifurcation point, and for constant angular momentum, the solutions that minimize the kinetic energy are the *non*-axially symmetric Jacobi ellipsoids instead of the Maclaurin spheroids.

References

- Alexander, Victoria N. (2011). The Biologist's Mistress: Rethinking Self-Organization in Art, Literature, and Nature. Litchfield Park, AZ: Emergent Publications. ISBN 0-9842165-5-3
- Lakhani, Archana; Banerjee, A; Chaddah, P; Chen, X; Ramanujan, R V (2012). "Magnetic glass in shape memory alloy: Ni45Co5Mn38Sn12". Journal of Physics: Condensed Matter. 24 (38): 386004. Bibcode:2012JPCM...24L6004L. ISSN 0953-8984. arXiv:1206.2024. doi:10.1088/0953-8984/24/38/386004
- Blundell, Stephen J.; Katherine M. Blundell (2008). Concepts in Thermal Physics. Oxford University Press. ISBN 978-0-19-856770-7
- Imry, Y.; Wortis, M. (1979). "Influence of quenched impurities on first-order phase transitions". Phys. Rev. B. 19 (7): 3580–3585. Bibcode:1979PhRvB..19.3580I. doi:10.1103/physrevb.19.3580

- Ivancevic, Vladimir G.; Ivancevic, Tijiana, T. (2008). Complex Nonlinearity. Berlin: Springer. pp. 176–177. 978-3-540-79357-1. Retrieved 12 October 2014
- YashRoy, R.C. (1987). "13-C NMR studies of lipid fatty acyl chains of chloroplast membranes". Indian Journal of Biochemistry and Biophysics. 24 (6): 177–178
- A. D. McNaught and A. Wilkinson (ed.). "Compendium of Chemical Terminology". IUPAC. ISBN 0-86542-684-8. Retrieved 2007-10-23
- Leonard, F.; Delamotte, B. (2015). "Critical exponents can be different on the two sides of a transition". Phys. Rev. Lett. 115: 200601. Bibcode:2015PhRvL.115t0601L. arXiv:1508.07852. doi:10.1103/PhysRevLett.115.200601
- Mora, Thierry; Bialek, William (2011). "Are biological systems poised at criticality?". Journal of Statistical Physics. 144 (2): 268–302. Bibcode:2011JSP...144..268M. arXiv:1012.2242. doi:10.1007/s10955-011-0229-4
- Luisi, Pier L. (2006). The Emergence of Life: From Chemical Origins to Synthetic Biology. Cambridge, England: Cambridge University Press. p. 119. ISBN 0521821177
- Steven Weinberg. "A Designer Universe?". Retrieved 2008-07-14. A version of the original quote from address at the Conference on Cosmic Design, American Association for the Advancement of Science, Washington, D.C. in April 1999
- Fu, Wai-Tat; Kannampallil, Thomas; Kang, Ruogu; He, Jibo (2010), "Semantic Imitation in Social Tagging", ACM Transactions on Computer-Human Interaction, 17 (3): 1–37, doi:10.1145/1806923.1806926
- Liouville, J. (1834). "Sur la figure d'une masse fluide homogène, en équilibre et douée d'un mouvement de rotation". Journal de l'École Polytechnique (14): 289–296
- Daniel Charles Mattis (1994). The Many-Body Problem: Encyclopaedia of Exactly Solved Models in One Dimension. World Scientific. p. 340. ISBN 981-02-1476-6
- Anderson, P.W. (1972). "More is Different" (PDF). Science. 177 (4047): 393–396. Bibcode:1972Sci...177..393A. PMID 17796623. doi:10.1126/science.177.4047.393

Theories Related to Condensed Matter Physics

4

Various theories related to condensed matter physics include Bloch wave, Ginzburg–Landau theory, Fermi gas, etc. Fermi gas contains a large number of fermions and is a phase of matter. The topics discussed in the chapter are of great importance to broaden the existing knowledge on condensed matter physics.

Quantum Field Theory

In theoretical physics, quantum field theory (QFT) is the theoretical framework for constructing quantum mechanical models of subatomic particles in particle physics and quasiparticles in condensed matter physics. QFT treats particles as excited states of the underlying physical field, so these are called field quanta.

In quantum field theory, quantum mechanical interactions among particles are described by interaction terms among the corresponding underlying quantum fields. These interactions are conveniently visualized by Feynman diagrams, which are a formal tool of relativistically covariant perturbation theory, serving to evaluate particle processes.

History

Even though QFT is an unavoidable consequence of the reconciliation of quantum mechanics with special relativity (Weinberg (1995)), historically, it emerged in the 1920s with the quantization of the electromagnetic field (the quantization being based on an analogy of the eigenmode expansion of a vibrating string with fixed endpoints).

Early Development

Max Born (1882–1970), one of the founders of quantum field theory.

He is also known for the Born rule that introduced the probabilistic interpretation in quantum mechanics. He received the 1954 Nobel Prize in Physics together with Walther Bothe.

The first achievement of quantum field theory, namely quantum electrodynamics (QED), is "still the paradigmatic example of a successful quantum field theory" (Weinberg (1995)). Ordinarily, quantum mechanics (QM) cannot give an account of photons which constitute the prime case of relativistic 'particles'. Since photons have rest mass zero, and correspondingly travel in the vacuum at the speed *c*, a non-relativistic theory such as ordinary QM cannot give even an approximate description. Photons are implicit in the emission and absorption processes which have to be postulated; for instance, when one of an atom's electrons makes a transition between energy levels. The formalism of QFT is needed for an explicit description of photons. In fact most topics in the early development of quantum theory (the so-called old quantum theory, 1900–25) were related to the interaction of radiation and matter and thus should be treated by quantum field theoretical methods. However, quantum mechanics as formulated by Dirac, Heisenberg, and Schrödinger in 1926–27 started from atomic spectra and did not focus much on problems of radiation.

As soon as the conceptual framework of quantum mechanics was developed, a small group of theoreticians tried to extend quantum methods to electromagnetic fields. A good example is the famous paper by Born, Jordan & Heisenberg (1926). (P. Jordan was especially acquainted with the literature on light quanta and made seminal contributions to QFT.) The basic idea was that in QFT the electromagnetic field should be represented by matrices in the same way that position and momentum were represented in QM by matrices (matrix mechanics oscillator operators). The ideas of QM were thus extended to systems having an infinite number of degrees of freedom, so an infinite array of quantum oscillators.

The inception of QFT is usually considered to be Dirac's famous 1927 paper on "The quantum theory of the emission and absorption of radiation". Here Dirac coined the name "quantum electrodynamics" (QED) for the part of QFT that was developed first. Dirac supplied a systematic procedure for transferring the characteristic quantum phenomenon of discreteness of physical quantities from the quantum-mechanical treatment of particles to a corresponding treatment of fields. Employing the theory of the quantum harmonic oscillator, Dirac gave a theoretical description of how photons appear in the quantization of the electromagnetic radiation field. Later, Dirac's procedure became a model for the quantization of other fields as well. These first approaches to QFT were further developed during the following three years. P. Jordan introduced creation and annihilation operators for fields obeying Fermi–Dirac statistics. These differ from the corresponding operators for Bose–Einstein statistics in that the former satisfy *anti-commutation relations* while the latter satisfy commutation relations.

The methods of QFT could be applied to derive equations resulting from the quantum-mechanical (field-like) treatment of particles, e.g. the Dirac equation, the Klein–Gordon equation and the Maxwell equations. Schweber points out that the idea and procedure of second quantization goes back to Jordan, in a number of papers from 1927, while the expression itself was coined by Dirac. Some difficult problems concerning commutation relations, statistics, and Lorentz invariance were eventually solved. The first comprehensive account of a general theory of quantum fields, in particular, the method of canonical quantization, was presented by Heisenberg & Pauli in 1929–30. Whereas Jordan's second quantization procedure applied to the coefficients of the normal modes of the field, Heisenberg & Pauli started with the fields themselves and subjected them to the canonical procedure. Heisenberg and Pauli thus established the basic structure of QFT as presented in modern introductions to QFT. Fermi and Dirac, as well as Fock and Podolsky, presented different formulations which played a heuristic role in the following years.

Quantum electrodynamics rests on two pillars, see e.g., the short and lucid "Historical Introduction" of Scharf (2014). The first pillar is the quantization of the electromagnetic field, i.e., it is about photons as the quantized excitations or 'quanta' of the electromagnetic field. As Weinberg points out the "photon is the only particle that was known as a field before it was detected as a particle" so that it is natural that QED began with the analysis of the radiation field. The second pillar of QED consists of the relativistic theory of the electron, centered on the Dirac equation.

The Problem of Infinities

The Emergence of Infinities

Pascual Jordan (1902–1980), doctoral student of Max Born, was a pioneer in quantum field theory, coauthoring a number of seminal papers with Born and Heisenberg.

Jordan algebras were introduced by him to formalize the notion of an algebra of observables in quantum mechanics. He was awarded the Max Planck medal 1954.

Quantum field theory started with a theoretical framework that was built in analogy to quantum mechanics. Although there was no unique and fully developed theory, quantum field theoretical tools could be applied to concrete processes. Examples are the scattering of radiation by free electrons, Compton scattering, the collision between relativistic electrons or the production of electron-positron pairs by photons. Calculations to the first order of approximation were quite successful, but most people working in the field thought that QFT still had to undergo a major change. On the one side, some calculations of effects for cosmic rays clearly differed from measurements. On the other side and, from a theoretical point of view more threatening, calculations of higher orders of the perturbation series led to infinite results. The self-energy of the electron as well as vacuum fluctuations of the electromagnetic field seemed to be infinite. The perturbation expansions did not converge to a finite sum and even most individual terms were divergent.

The various forms of infinities suggested that the divergences were more than failures of specific calculations. Many physicists tried to avoid the divergences by formal tricks (truncating the integrals at some value of momentum, or even ignoring infinite terms) but such rules were not reliable, violated the requirements of relativity and were not considered as satisfactory. Others came up with the first ideas for coping with infinities by a redefinition of the parameters of the theory and using a measured finite value, for example of the charge of the electron, instead of the infinite 'bare' value. This process is called renormalization.

From the point of view of the philosophy of science, it is remarkable that these divergences did not give enough reason to discard the theory. The years from 1930 to the beginning of World War II were characterized by a variety of attitudes towards QFT. Some physicists tried to circumvent the infinities by more-or-less arbitrary prescriptions, others worked on transformations and improvements of the theoretical framework. Most of the theoreticians believed that QED would break down at high energies. There was also a considerable number of proposals in favor of alternative approaches. These proposals included changes in the basic concepts e.g. negative probabilities and interactions at a distance instead of a field theoretical approach, and a methodological change to phenomenological methods that focusses on relations between observable quantities without an analysis of the microphysical details of the interaction, the so-called S-matrix theory where the basic elements are amplitudes for various scattering processes.

Despite the feeling that QFT was imperfect and lacking rigor, its methods were extended to new areas of applications. In 1933 Fermi's theory of the beta decay started with conceptions describing the emission and absorption of photons, transferred them to beta radiation and analyzed the creation and annihilation of electrons and neutrinos described by the weak interaction. Further applications of QFT outside of quantum electrodynamics succeeded in nuclear physics with the strong interaction. In 1934 Pauli & Weisskopf showed that a new type of field (the scalar field), described by the Klein–Gordon equation, could be quantized. This is another example of second quantization. This new theory for matter fields could be applied a decade later when new particles, pions, were detected.

The Taming of Infinities

Werner Heisenberg (1901–1976), doctoral student of Arnold Sommerfeld, was one of the founding fathers of quantum mechanics and QFT.

In particular, he introduced the version of quantum mechanics known as matrix mechanics, but is now more known for the Heisenberg uncertainty relations. He was awarded the Nobel prize in physics 1932.

After the end of World War II more reliable and effective methods for dealing with infinities in QFT were developed, namely coherent and systematic rules for performing relativistic field theoretical calculations, and a general renormalization theory. At three famous conferences, the Shelter Island Conference 1947, the Pocono Conference 1948, and the 1949 Oldstone Conference, developments in theoretical physics were confronted with relevant new experimental results. In the late forties, there were two different ways to address the problem of divergences. One of these was discovered by Richard Feynman, the other one (based on an operator formalism) by Julian Schwinger and, independently, by Shin'ichirō Tomonaga.

In 1949, Freeman Dyson showed that the two approaches are in fact equivalent and fit into an elegant field-theoretic framework. Thus, Freeman Dyson, Feynman, Schwinger, and Tomonaga became the inventors of renormalization theory. The most spectacular successes of renormalization theory were the calculations of the anomalous magnetic moment of the electron and the Lamb shift in the spectrum of hydrogen. These successes were so outstanding because the theoretical results were in better agreement with high-precision experiments than anything in physics encountered before. Nevertheless, mathematical problems lingered on and prompted a search for rigorous formulations.

The rationale behind renormalization is to avoid divergences that appear in physical predictions by shifting them into a part of the theory where they do not influence empirical statements. Dyson could show that a rescaling of charge and mass ('renormalization') is sufficient to remove all divergences in QED consistently, to all orders of perturbation theory. A QFT is called renormalizable if all infinities can be absorbed into a redefinition of a *finite number* of coupling constants and masses. A consequence for QED is that the physical charge and mass of the electron must be measured and cannot be computed from first principles.

Perturbation theory yields well-defined predictions only in renormalizable quantum field theories; luckily, QED, the first fully developed QFT, belonged to this class of renormalizable theories. There are various technical procedures to renormalize a theory. One way is to cut off the integrals in the calculations at a certain value Λ of the momentum which is large but finite. This cut-off procedure is successful if, after taking the limit $\Lambda \to \infty$, the resulting quantities are independent of Λ.

Richard Feynman (1918–1988) His 1945 PhD thesis developed the path integral formulation of ordinary quantum mechanics. This was later generalized to field theory.

Feynman's formulation of QED is of special interest from a philosophical point of view. His so-called space-time approach is visualized by the celebrated Feynman diagrams that look like depicting paths of particles. Feynman's method of calculating scattering amplitudes is based on the functional integral formulation of field theory. A set of graphical rules can be derived so that the probability of a specific scattering process can be calculated by drawing a diagram of that process and then using that diagram to write down the precise mathematical expressions for calculating its amplitude in relativistically covariant perturbation theory.

The diagrams provide an effective way to organize and visualize the various terms in the perturbation series, and they naturally account for the flow of electrons and photons during the scattering process. External lines in the diagrams represent incoming and outgoing particles, internal lines are connected with virtual particles and vertices with interactions. Each of these graphical elements is associated with mathematical expressions that contribute to the amplitude of the respective process. The diagrams are part of Feynman's very efficient and elegant algorithm for computing the probability of scattering processes.

The idea of particles traveling from one point to another was heuristically useful in constructing the theory. This heuristics, based on Huygens' principle, is useful for concrete calculations and actually give the correct particle propagators as derived more rigorously. Nevertheless, an analysis of the theoretical justification of the space-time approach shows that its success does not imply that particle paths need be taken seriously. General arguments against a particle interpretation of QFT clearly exclude that the diagrams represent actual paths of particles in the interaction area. Feynman himself was not particularly interested in ontological questions.

The Golden Age: Gauge Theory and the Standard Model

Chen-Ning Yang (b. 1922), co-inventor of nonabelian gauge field theories

In 1933, Enrico Fermi had already established that the creation, annihilation and transmutation of particles in the weak interaction beta decay could best be described in QFT, specifically his quartic fermion interaction. As a result, field theory had become a prospective tool for other particle interactions. In the beginning of the 1950s, QED had become a reliable theory which no longer counted as preliminary. However, it took two decades from writing down the first equations until QFT could be applied successfully to important physical problems in a systematic way.

Murray Gell-Mann (b. 1929) articulator and pioneer of group symmetry in QFT

The theories explored relied on—indeed, were virtually fully specified by—a rich variety of symmetries pioneered and articulated by Murray Gell-Mann. The new developments made it possible to apply QFT to new particles and new interactions and fully explain their structure.

In the following decades, QFT was extended to well-describe not only the electromagnetic force but also weak and strong interaction so that new Lagrangians were found which contain new classes of particles or quantum fields. The search still continues for a more comprehensive theory of matter and energy, a *unified theory of all interactions.*

Yoichiro Nambu (1921–2015), co-discoverer of field theoretic spontaneous symmetry breaking

The new focus on symmetry led to the triumph of non-Abelian gauge theories (the development of such theories was pioneered in 1954–60 with the work of Yang and Mills) and spontaneous symmetry breaking (by Yoichiro Nambu). Today, there are reliable theories of the strong, weak, and electromagnetic interactions of elementary particles which have an analogous structure to QED: They are the dominant framework of particle physics.

A combined renormalizable theory associated with the gauge group $SU(3) \times SU(2) \times U(1)$ is dubbed the *standard model of elementary particle physics* (even though it is a full theory, and not just a model) and was assembled by Sheldon Glashow, Steven Weinberg and Abdus Salam in 1959–67, and Frank Wilczek, David Gross and David Politzer in 1973, on the basis of conceptual breakthroughs by Peter Higgs, François Englert, Robert Brout, Martin Veltman, and Gerard 't Hooft.

Gerard 't Hooft (b.1946) proved gauge field theories are renormalizable

According to the standard model, there are, on the one hand, six types of leptons (e.g. the electron and its neutrino) and six types of quarks, where the members of both groups are all fermions with spin 1/2. On the other hand, there are spin 1 particles (thus bosons) that mediate the interaction between elementary particles and the fundamental forces, namely the photon for electromagnetic interaction, two W and one Z-boson for weak interaction, and the gluons for strong interaction. The linchpin of the symmetry breaking mechanism of the theory is the spin 0 Higgs boson, discovered 40 years after its prediction.

Renormalization Group

Parallel breakthroughs in the understanding of phase transitions in condensed matter physics led to novel insights based on the renormalization group. They emerged in the work of Leo Kadanoff (1966) and Kenneth Geddes Wilson & Michael Fisher (1972)—extending the work of Ernst Stueckelberg–André Petermann (1953) and Murray Gell-Mann–Francis Low (1954)—which led to the seminal reformulation of quantum field theory by Kenneth Geddes Wilson in 1975. This reformulation provided insights into the evolution of effective field theories with scale, which classified all field theories, renormalizable or not (cf. subsequent section). The remarkable conclusion is that, in general, most observables are *"irrelevant"*, i.e., the macroscopic physics is *dominated by only a few observables* in most systems.

During the same period, Kadanoff (1969) introduced an operator algebra formalism for the two-dimensional Ising model, a widely studied mathematical model of ferromagnetism in statistical physics. This development suggested that quantum field theory describes its scaling limit. Later, there developed the idea that a finite number of generating operators could represent all the correlation functions of the Ising model.

Conformal Field Theory

The existence of a much stronger symmetry for the scaling limit of two-dimensional critical systems was suggested by Alexander Belavin, Alexander Polyakov and Alexander Zamolodchikov in 1984, which eventually led to the development of conformal field theory, a special case of quantum field theory, which is presently utilized in different areas of particle physics and condensed matter physics.

Historiography

The first chapter in Weinberg (1995) is a very good short description of the earlier history of QFT. A detailed account of the historical development of QFT can be found in Schweber (1994).

Varieties of Approaches

Most theories in standard particle physics are formulated as *relativistic quantum field theories*, such as QED, QCD, and the Standard Model. QED, the quantum field-theoretic description of the electromagnetic field, approximately reproduces Maxwell's theory of electrodynamics in the low-energy limit, with small non-linear corrections to the Maxwell equations required due to virtual electron–positron pairs.

Perturbative and Non-perturbative Approaches

In the perturbative approach to quantum field theory, the full field interaction terms are approximated as a perturbative expansion in the number of particles involved. Each term in the expansion can be thought of as forces between particles being mediated by other particles. In QED, the electromagnetic force between two electrons is caused by an exchange of photons. Similarly, intermediate vector bosons mediate the weak force and gluons mediate the strong force in QCD. The notion of a force-mediating particle comes from perturbation theory, and does not make sense in the context of non-perturbative approaches to QFT, such as with bound states.

QFT and Gravity

There is currently no complete quantum theory of the remaining fundamental force, gravity. Many of the proposed theories to describe gravity as a QFT postulate the existence of a graviton particle that mediates the gravitational force. Presumably, the as yet unknown correct quantum field-theoretic treatment of the gravitational field will behave like Einstein's general theory of relativity in the low-energy limit. Quantum field theory of the fundamental forces itself has been postulated to be the low-energy effective field theory limit of a more fundamental theory such as superstring theory.

Definition

Quantum electrodynamics (QED) has one electron field and one photon field; quantum chromodynamics (QCD) has one field for each type of quark; and, in condensed matter, there is an atomic displacement field that gives rise to phonon particles. Edward Witten describes QFT as "by far" the most difficult theory in modern physics – "so difficult that nobody fully believed it for 25 years."

Dynamics

Ordinary quantum mechanical systems have a fixed number of particles, with each particle having a finite number of degrees of freedom. In contrast, the excited states of a quantum field can represent any number of particles. This makes quantum field theories especially useful for describing systems where the particle count/number may change over time, a crucial feature of relativistic dynamics. A QFT is thus an organized infinite array of oscillators.

States

QFT interaction terms are similar in spirit to those between charges with electric and magnetic fields in Maxwell's equations. However, unlike the classical fields of Maxwell's theory, fields in QFT generally exist in quantum superpositions of states and are subject to the laws of quantum mechanics.

Because the fields are continuous quantities over space, there exist excited states with arbitrarily large numbers of particles in them, providing QFT systems with effectively an infinite number of degrees of freedom. Infinite degrees of freedom can easily lead to divergences of calculated quantities (e.g., the quantities become infinite). Techniques such as renormalization of QFT parameters or discretization of spacetime, as in lattice QCD, are often used to avoid such infinities so as to yield physically plausible results.

Fields and Radiation

The gravitational field and the electromagnetic field are the only two fundamental fields in nature that have infinite range and a corresponding classical low-energy limit, which greatly diminishes and hides their "particle-like" excitations. Albert Einstein in 1905, attributed "particle-like" and discrete exchanges of momenta and energy, characteristic of "field quanta", to the electromagnetic field. Originally, his principal motivation was to explain the thermodynamics of radiation. Although the photoelectric effect and Compton scattering strongly suggest the existence of the photon, it might alternatively be explained by a mere quantization of emission; more definitive evidence of the quantum nature of radiation is now taken up into modern quantum optics as in the antibunching effect.

Principles

Classical and Quantum Fields

A classical field is a function defined over some region of space and time. Two physical phenomena which are described by classical fields are Newtonian gravitation, described by Newtonian gravitational field g(x, *t*), and classical electromagnetism, described by the electric and magnetic fields E(x, *t*) and B(x, *t*). Because such fields can in principle take on distinct values at each point in space, they are said to have infinite degrees of freedom.

Classical field theory does not, however, account for the quantum-mechanical aspects of such physical phenomena. For instance, it is known from quantum mechanics that certain aspects of electromagnetism involve discrete particles—photons—rather than continuous fields. The business of *quantum* field theory is to write down a field that is, like a classical field, a function defined over space and time, but which also accommodates the observations of quantum mechanics. This is a *quantum field*.

To write down such a quantum field, one promotes the infinity of classical oscillators representing the modes of the classical fields to quantum harmonic oscillators. They thus become operator-valued functions (actually, distributions). (In its most general formulation, quantum mechanics is a theory of abstract operators (observables) acting on an abstract state space (Hilbert space), where the observables represent physically observable quantities and the state space represents the possible states of the system under study. For instance, the fundamental observables associated with the motion of a single quantum mechanical particle are the position and momentum operators $\hat{x}$ and $\hat{p}$. Field theory, by sharp contrast, treats x as a label, an index of the field rather than as an operator.)

There are two common ways of handling a quantum field: canonical quantization and the path integral formalism. The latter of these is pursued in this article.

Lagrangian Formalism

Quantum field theory relies on the Lagrangian formalism from classical field theory. This formalism is analogous to the Lagrangian formalism used in classical mechanics to solve for the motion of a particle under the influence of a field. In classical field theory, one writes down a Lagrangian density, $\mathcal{L}$, involving a field, φ(x,t), and possibly its first derivatives (∂φ/∂t and ∇φ), and then applies a field-theoretic form of the Euler–Lagrange equation. Writing coordinates (t, x) = (x^0, x^1, x^2, x^3) = xμ, this form of the Euler–Lagrange equation is

$$\frac{\partial}{\partial x^{\mu}}\left[\frac{\partial \mathcal{L}}{\partial(\partial\varphi / \partial x^{\mu})}\right]-\frac{\partial \mathcal{L}}{\partial \varphi}=0,$$

where a sum over μ is performed according to the rules of Einstein notation.

By solving this equation, one arrives at the "equations of motion" of the field. For example, if one begins with the Lagrangian density

$$\mathcal{L}(\varphi,\nabla\varphi)=-\rho(t,\mathbf{x})\varphi(t,\mathbf{x})-\frac{1}{8\pi G}|\nabla\varphi|^{2},$$

and then applies the Euler–Lagrange equation, one obtains the equation of motion

$$4\pi G\rho(t,\mathbf{x})=\nabla^{2}\varphi.$$

This equation is Newton's law of universal gravitation, expressed in differential form in terms of the gravitational potential φ(*t*, x) and the mass density ρ(*t*, x). Despite the nomenclature, the "field" under study is the gravitational potential, φ, rather than the gravitational field, g. Similarly, when classical field theory is used to study electromagnetism, the "field" of interest is the electromagnetic four-potential (*V*/*c*, A), rather than the electric and magnetic fields E and B.

Quantum field theory uses this same Lagrangian procedure to determine the equations of motion for quantum fields. These equations of motion are then supplemented by commutation relations derived from the canonical quantization procedure described below, thereby incorporating quantum mechanical effects into the behavior of the field.

Single- and Many-particle Quantum Mechanics

In non-relativistic quantum mechanics, a particle (such as an electron or proton) is described by a complex wavefunction, $\psi(x, t)$, whose time-evolution is governed by the Schrödinger equation:

$$-\frac{\hbar^{2}}{2m}\frac{\partial^{2}}{\partial x^{2}}\psi(x,t)+V(x)\psi(x,t)=i\hbar\frac{\partial}{\partial t}\psi(x,t).$$

Here m is the particle's mass and $V(x)$ is the applied potential. Physical information about the behavior of the particle is extracted from the wavefunction by constructing expected values for various quantities; for example, the expected value of the particle's position is given by integrating $\psi^*(x)\, x\, \psi(x)$ over all space, and the expected value of the particle's momentum is found by integrating $-i\hbar\psi^*(x)d\psi/dx$. The quantity $\psi^*(x)\psi(x)$ is itself in the Copenhagen interpretation of quantum mechanics interpreted as a probability density function. This treatment of quantum mechanics, where a particle's wavefunction evolves against a classical background potential $V(x)$, is sometimes called *first quantization*.

This description of quantum mechanics can be extended to describe the behavior of multiple particles, so long as the number and the type of particles remain fixed. The particles are described by a wavefunction $\psi(x_1, x_2, ..., x_N, t)$, which is governed by an extended version of the Schrödinger equation.

Often one is interested in the case where N particles are all of the same type (for example, the 18 electrons orbiting a neutral argon nucleus). As described on identical particles, this implies that the state of the entire system must be either symmetric (bosons) or antisymmetric (fermions) when the coordinates of its constituent particles are exchanged. This is achieved by using a Slater determinant as the wavefunction of a fermionic system (and a Slater permanent for a bosonic system), which is equivalent to an element of the symmetric or antisymmetric subspace of a tensor product.

For example, the general quantum state of a system of N bosons is written as

$$|\phi_1 \cdots \phi_N\rangle = \sqrt{\frac{\prod_j N_j!}{N!}} \sum_{p\in S_N} |\phi_{p(1)}\rangle \otimes \cdots \otimes |\phi_{p(N)}\rangle,$$

where $|\phi_i\rangle$ are the single-particle states, Nj is the number of particles occupying state j, and the sum is taken over all possible permutations p acting on N elements. In general, this is a sum of N! (N factorial) distinct terms. $\sqrt{\frac{\prod_j N_j!}{N!}}$ is a normalizing factor.

There are several shortcomings to the above description of quantum mechanics, which are addressed by quantum field theory. First, it is unclear how to extend quantum mechanics to include the effects of special relativity. Attempted replacements for the Schrödinger equation, such as the Klein–Gordon equation or the Dirac equation, have many unsatisfactory qualities; for instance, they possess energy eigenvalues that extend to $-\infty$, so that there seems to be no easy definition of a ground state. It turns out that such inconsistencies arise from relativistic wavefunctions not having a well-defined probabilistic interpretation in position space, as probability conservation is not a relativistically covariant concept. The second shortcoming, related to the first, is that in quantum mechanics there is no mechanism to describe particle creation and annihilation; this is crucial for describing phenomena such as pair production, which result from the conversion between mass and energy according to the relativistic relation $E = mc^2$.

Second Quantization

In this section, we will describe a method for constructing a quantum field theory called second quantization. This basically involves choosing a way to index the quantum mechanical degrees of freedom in the space of multiple identical-particle states. It is based on the Hamiltonian formulation of quantum mechanics.

Several other approaches exist, such as the Feynman path integral, which uses a Lagrangian formulation.

Bosons

For simplicity, we will first discuss second quantization for bosons, which form perfectly symmetric quantum states. Let us denote the mutually orthogonal single-particle states which are possible in the system by $|\phi_1\rangle, |\phi_2\rangle, |\phi_3\rangle$ and so on. For example, the 3-particle state with one particle in state $|\phi_1\rangle$ and two in state $|\phi_2\rangle$ is

$$\frac{1}{\sqrt{3}}\left[|\phi_1\rangle|\phi_2\rangle|\phi_2\rangle+|\phi_2\rangle|\phi_1\rangle|\phi_2\rangle+|\phi_2\rangle|\phi_2\rangle|\phi_1\rangle\right].$$

The first step in second quantization is to express such quantum states in terms of occupation numbers, by listing the number of particles occupying each of the single-particle states $|\phi_1\rangle, |\phi_2\rangle$ etc. This is simply another way of labelling the states. For instance, the above 3-particle state is denoted as

$$|1,2,0,0,0,\ldots\rangle.$$

An N-particle state belongs to a space of states describing systems of N particles. The next step is to combine the individual N-particle state spaces into an extended state space, known as Fock space, which can describe systems of any number of particles. This is composed of the state space of a system with no particles (the so-called vacuum state, written as $|0\rangle$), plus the state space of a 1-particle system, plus the state space of a 2-particle system, and so forth. States describing a definite number of particles are known as Fock states: a general element of Fock space will be a linear combination of Fock states. There is a one-to-one correspondence between the occupation number representation and valid boson states in the Fock space.

At this point, the quantum mechanical system has become a quantum field in the sense we described above. The field's elementary degrees of freedom are the occupation numbers, and each occupation number is indexed by a number j indicating which of the single-particle states $|\phi_1\rangle, |\phi_2\rangle, \ldots, |\phi_j\rangle, \ldots$ it refers to:

$$|N_1, N_2, N_3, \ldots, N \ ,\ldots\rangle.$$

The properties of this quantum field can be explored by defining creation and annihilation operators, which add and subtract particles. They are analogous to ladder operators in the quantum harmonic oscillator problem, which added and subtracted energy quanta. However, these operators literally create and annihilate particles of a given quantum state. The bosonic annihilation operator a_2 and creation operator $a_2^{\dagger}$ are easily defined in the occupation number representation as having the following effects:

$$a_2 \mid N_1, N_2, N_3, \ldots\rangle = \sqrt{N_2} \mid N_1, (N_2 - 1), N_3, \ldots\rangle,$$

$$a_2^\dagger \mid N_1, N_2, N_3, \ldots\rangle = \sqrt{N_2 + 1} \mid N_1, (N_2 + 1), N_3, \ldots\rangle.$$

It can be shown that these are operators in the usual quantum mechanical sense, i.e. linear operators acting on the Fock space. Furthermore, they are indeed Hermitian conjugates, which justifies the way we have written them. They can be shown to obey the commutation relation

$$\left[a_i, a_j\right] = 0 \quad , \quad \left[a_i^\dagger, a_j^\dagger\right] = 0 \quad , \quad \left[a_i, a_j^\dagger\right] = \delta_{ij},$$

where δ stands for the Kronecker delta. These are precisely the relations obeyed by the ladder operators for an infinite set of independent quantum harmonic oscillators, one for each single-particle state. Adding or removing bosons from each state is, therefore, analogous to exciting or de-exciting a quantum of energy in a harmonic oscillator.

Applying an annihilation operator a_k followed by its corresponding creation operator $a_k^\dagger$ returns the number N_k of particles in the kth single-particle eigenstate:

$$a_k^\dagger a_k \mid \ldots, N_k, \ldots\rangle = N_k \mid \ldots, N_k, \ldots\rangle.$$

The combination of operators $a_k^\dagger a_k$ is known as the number operator for the kth eigenstate.

The Hamiltonian operator of the quantum field (which, through the Schrödinger equation, determines its dynamics) can be written in terms of creation and annihilation operators. For instance, for a field of free (non-interacting) bosons, the total energy of the field is found by summing the energies of the bosons in each energy eigenstate. If the kth single-particle energy eigenstate has energy E_k and there are bosons in this state, then the total energy of these bosons is $E_k N_k$. The energy in the entire field is then a sum over k:

$$E_{tot} = \sum_k E_k N_k$$

This can be turned into the Hamiltonian operator of the field by replacing with the corresponding number operator, $a_k^\dagger a_k$. This yields

$$H = \sum_k E_k a_k^\dagger a_k.$$

Fermions

It turns out that a different definition of creation and annihilation must be used for describing fermions. According to the Pauli exclusion principle, fermions cannot share quantum states, so their occupation numbers Ni can only take on the value 0 or 1. The fermionic annihilation operators c and creation operators $c^\dagger$ are defined by their actions on a Fock state thus

$$c_j \mid N_1, N_2, \ldots, N_j = 0, \ldots\rangle = 0$$

$$c_j \mid N_1, N_2, \ldots, N_j = 1, \ldots \rangle = (-1)^{(N_1 + \cdots + N_{j-1})} \mid N_1, N_2, \ldots, N_j = 0, \ldots \rangle$$

$$c_j^{\dagger} \mid N_1, N_2, \ldots, N_j = 0, \ldots \rangle = (-1)^{(N_1 + \cdots + N_{j-1})} \mid N_1, N_2, \ldots, N_j = 1, \ldots \rangle$$

$$c_j^{\dagger} \mid N_1, N_2, \ldots, N_j = 1, \ldots \rangle = 0.$$

These obey an anticommutation relation:

$$\left\{c_i, c_j\right\} = 0 \quad , \quad \left\{c_i^{\dagger}, c_j^{\dagger}\right\} = 0 \quad , \quad \left\{c_i, c_j^{\dagger}\right\} = \delta_{ij}.$$

One may notice from this that applying a fermionic creation operator twice gives zero, so it is impossible for the particles to share single-particle states, in accordance with the exclusion principle.

Field Operators

We have previously mentioned that there can be more than one way of indexing the degrees of freedom in a quantum field. Second quantization indexes the field by enumerating the single-particle quantum states. However, as we have discussed, it is more natural to think about a "field", such as the electromagnetic field, as a set of degrees of freedom indexed by position.

To this end, we can define *field operators* that create or destroy a particle at a particular point in space. In particle physics, these operators turn out to be more convenient to work with, because they make it easier to formulate theories that satisfy the demands of relativity.

Single-particle states are usually enumerated in terms of their momenta (as in the particle in a box problem.) We can construct field operators by applying the Fourier transform to the creation and annihilation operators for these states. For example, the bosonic field annihilation operator $\phi(\mathbf{r})$ is

$$\phi(\mathbf{r}) \overset{def}{=} \sum_j e^{i\mathbf{k}_j \cdot \mathbf{r}} a_j.$$

The bosonic field operators obey the commutation relation

$$\left[\phi(\mathbf{r}), \phi(\mathbf{r}')\right] = 0 \quad , \quad \left[\phi^{\dagger}(\mathbf{r}), \phi^{\dagger}(\mathbf{r}')\right] = 0 \quad , \quad \left[\phi(\mathbf{r}), \phi^{\dagger}(\mathbf{r}')\right] = \delta^3(\mathbf{r} - \mathbf{r}')$$

where $\delta(x)$ stands for the Dirac delta function. As before, the fermionic relations are the same, with the commutators replaced by anticommutators.

The field operator is not the same thing as a single-particle wavefunction. The former is an operator acting on the Fock space, and the latter is a quantum-mechanical amplitude for finding a particle in some position. However, they are closely related and are indeed commonly denoted with the same symbol. If we have a Hamiltonian with a space representation, say

$$H = -\frac{\hbar^2}{2m} \sum_i \nabla_i^2 + \sum_{i<j} U(\mid \mathbf{r}_i - \mathbf{r}_j \mid)$$

where the indices i and j run over all particles, then the field theory Hamiltonian (in the non-relativistic limit and for negligible self-interactions) is

$$\mathrm{H} = -\frac{\hbar^2}{2m}\int d^3r\,\phi^\dagger(\mathbf{r})\nabla^2\phi(\mathbf{r}) + \frac{1}{2}\int d^3r\int d^3r'\,\phi^\dagger(\mathbf{r})\phi^\dagger(\mathbf{r}')U(|\mathbf{r}-\mathbf{r}'|)\phi(\mathbf{r}')\phi(\mathbf{r}).$$

This looks remarkably like an expression for the expectation value of the energy, with ϕ playing the role of the wavefunction. This relationship between the field operators and wave functions makes it very easy to formulate field theories starting from space projected Hamiltonians.

Dynamics

Once the Hamiltonian operator is obtained as part of the canonical quantization process, the time dependence of the state is described with the Schrödinger equation, just as with other quantum theories. Alternatively, the Heisenberg picture can be used where the time dependence is in the operators rather than in the states.

Probability amplitudes of observables in such systems are quite hard to evaluate, an enterprise which has absorbed considerable ingenuity in the last three quarters of a century. In practice, most often, expectation values of operators are computed systematically through *covariant perturbation theory*, formulated through Feynman diagrams, but path integral computer simulations have also produced important results. Contemporary particle physics relies on extraordinarily accurate predictions of such techniques.

Implications

Unification of Fields and Particles

The "second quantization" procedure outlined takes a set of single-particle quantum states as a starting point. Sometimes, it is impossible to define such single-particle states, and one must proceed directly to quantum field theory. For example, a quantum theory of the electromagnetic field *must* be a quantum field theory, because it is impossible (for various reasons) to define a wavefunction for a single photon. In such situations, the quantum field theory can be constructed by examining the mechanical properties of the classical field and guessing the corresponding quantum theory. For free (non-interacting) quantum fields, the quantum field theories obtained in this way have the same properties as those obtained using second quantization, such as well-defined creation and annihilation operators obeying commutation or anticommutation relations.

Quantum field theory thus provides a unified framework for describing "field-like" objects (such as the electromagnetic field, whose excitations are photons) and "particle-like" objects (such as electrons, which are treated as excitations of an underlying electron field), so long as one can treat interactions as "perturbations" of free fields.

Physical Meaning of Particle Indistinguishability

The second quantization procedure relies crucially on the particles being identical. We would not have been able to construct a quantum field theory from a distinguishable many-particle system, because there would have been no way of separating and indexing the degrees of freedom.

Many physicists prefer to take the converse interpretation, which is that *quantum field theory explains what identical particles are.* In ordinary quantum mechanics, there is not much theoretical motivation for using symmetric (bosonic) or antisymmetric (fermionic) states, and the need for such states is simply regarded as an empirical fact. From the point of view of quantum field theory, particles are identical if and only if they are excitations of the same underlying quantum field. Thus, the question "why are all electrons identical?" arises from mistakenly regarding individual electrons as fundamental objects, when in fact it is only the electron field that is fundamental.

Particle Conservation and Non-conservation

During second quantization, we started with a Hamiltonian and state space describing a fixed number of particles (N), and ended with a Hamiltonian and state space for an arbitrary number of particles. Of course, in many common situations N is an important and perfectly well-defined quantity, e.g. if we are describing a gas of atoms sealed in a box. From the point of view of quantum field theory, such situations are described by quantum states that are eigenstates of the number operator N, which measures the total number of particles present. As with any quantum mechanical observable, N is conserved if it commutes with the Hamiltonian. In that case, the quantum state is trapped in the N-particle subspace of the total Fock space, and the situation could equally well be described by ordinary N-particle quantum mechanics. (Strictly speaking, this is only true in the noninteracting case or in the low energy density limit of renormalized quantum field theories)

For example, we can see that the free boson Hamiltonian described above conserves particle number. Whenever the Hamiltonian operates on a state, each particle destroyed by an annihilation operator a_k is immediately put back by the creation operator $a_k^\dagger$.

On the other hand, it is possible, and indeed common, to encounter quantum states that are not eigenstates of N, which do not have well-defined particle numbers. Such states are difficult or impossible to handle using ordinary quantum mechanics, but they can be easily described in quantum field theory as quantum superpositions of states having different values of N. For example, suppose we have a bosonic field whose particles can be created or destroyed by interactions with a fermionic field. The Hamiltonian of the combined system would be given by the Hamiltonians of the free boson and free fermion fields, plus a "potential energy" term such as

$$H_I = \sum_{k,q} V_q (a_q + a_{-q}^\dagger) c_{k+q}^\dagger c_k,$$

where $a_k^\dagger$ and a_k denotes the bosonic creation and annihilation operators, $c_k^\dagger$ and c_k denotes the fermionic creation and annihilation operators, and V_q is a parameter that describes the strength of the interaction. This "interaction term" describes processes in which a fermion in state k either absorbs or emits a boson, thereby being kicked into a different eigenstate $k+q$. (In fact, this type of Hamiltonian is used to describe the interaction between conduction electrons and phonons in metals. The interaction between electrons and photons is treated in a similar way, but is a little more complicated because the role of spin must be taken into account.) One thing to notice here is that even if we start out with a fixed number of bosons, we will typically end up with a superposition of states with different numbers of bosons at later times. The number of fermions, however, is conserved in this case.

In condensed matter physics, states with ill-defined particle numbers are particularly important for describing the various superfluids. Many of the defining characteristics of a superfluid arise from the notion that its quantum state is a superposition of states with different particle numbers. In addition, the concept of a coherent state (used to model the laser and the BCS ground state) refers to a state with an ill-defined particle number but a well-defined phase.

Associated Phenomena

Beyond the most general features of quantum field theories, special aspects such as renormalizability, gauge symmetry, and supersymmetry are outlined below.

Renormalization

Early in the history of quantum field theory, as detailed above, it was found that many seemingly innocuous calculations, such as the perturbative shift in the energy of an electron due to the presence of the electromagnetic field, yield infinite results. The reason is that the perturbation theory for the shift in an energy involves a sum over *all other energy levels, and there are infinitely many levels* at short distances, so that each gives a finite contribution which results in a divergent series.

Many of these problems are related to failures in classical electrodynamics that were identified but unsolved in the 19th century, and they basically stem from the fact that many of the supposedly "intrinsic" properties of an electron are tied to the electromagnetic field that it carries around with it. The energy carried by a single electron—its self-energy—is not simply the bare value, but also includes the energy contained in its electromagnetic field, its attendant cloud of photons. The energy in a field of a spherical source diverges in both classical and quantum mechanics, but as discovered by Weisskopf with help from Furry, in quantum mechanics *the divergence is much milder*, going *only as the logarithm* of the radius of the sphere.

The solution to the problem, presciently suggested by Stueckelberg, independently by Bethe after the crucial experiment by Lamb and Retherford (the Lamb–Retherford experiment), implemented at one loop by Schwinger, and systematically extended to all loops by Feynman and Dyson, with converging work by Tomonaga in isolated postwar Japan, comes from recognizing that all the infinities in the interactions of photons and electrons can be isolated into redefining a finite number of quantities in the equations by replacing them with the observed values: specifically the electron's mass and charge: this is called renormalization. The technique of renormalization recognizes that the problem is tractable and essentially purely mathematical; and that, physically, extremely short distances are at fault.

In order to define a theory on a continuum, one may first place a cutoff on the fields, by postulating that quanta cannot have energies above some extremely high value. This has the effect of replacing continuous space by a structure where very short wavelengths do not exist, as on a lattice. Lattices break rotational symmetry, and one of the crucial contributions made by Feynman, Pauli and Villars, and modernized by 't Hooft and Veltman, is a symmetry-preserving cutoff for perturbation theory (this process is called regularization). There is no known symmetrical cutoff outside of perturbation theory, so for rigorous or numerical work people often use an actual lattice.

On a lattice, every quantity is finite but depends on the spacing. When taking the limit to zero spacing, one makes sure that the physically observable quantities like the observed electron mass stay fixed, which means that the constants in the Lagrangian defining the theory depend on the spacing. By allowing the constants to vary with the lattice spacing, all the results at long distances become insensitive to the lattice, defining a continuum limit.

The renormalization procedure only works for a certain limited class of quantum field theories, called *renormalizable quantum field theories*. A theory is *perturbatively renormalizable* when the constants in the Lagrangian only diverge at worst as *logarithms* of the lattice spacing for very short spacings. The continuum limit is then well defined in perturbation theory, and even if it is not fully well defined non-perturbatively, the problems only show up at distance scales that are exponentially small in the inverse coupling for weak couplings. The Standard Model of particle physics is perturbatively renormalizable, and so are its component theories (quantum electrodynamics/electroweak theory and quantum chromodynamics). Of the three components, quantum electrodynamics is believed to not have a continuum limit by itself, while the asymptotically free *SU(2)* and *SU(3)* weak and strong color interactions are nonperturbatively well defined.

The renormalization group as developed along Wilson's breakthrough insights relates effective field theories at a given scale to such at contiguous scales. It thus describes how renormalizable theories emerge as the long distance low-energy effective field theory for any given high-energy theory. As a consequence, renormalizable theories are insensitive to the precise nature of the underlying high-energy short-distance phenomena (the macroscopic physics is dominated by only a few *"relevant" observables*). This is a blessing in practical terms, because it allows physicists to formulate low energy theories without detailed knowledge of high-energy phenomena. It is also a curse, because once a renormalizable theory such as the standard model is found to work, it provides very few clues to higher-energy processes.

The only way high-energy processes can be seen in the standard model is when they allow otherwise forbidden events, or else if they reveal predicted compelling quantitative relations among the coupling constants of the theories or models.

On account of renormalization, the couplings of QFT vary with scale, thereby confining quarks into hadrons, allowing the study of weakly-coupled quarks inside hadrons, and enabling speculation on ultra-high energy behavior.

Gauge Freedom

A gauge theory is a theory that admits a symmetry with a local parameter. For example, in every quantum theory, the global phase of the wave function is arbitrary and does not represent something physical. Consequently, the theory is invariant under a global change of phases (adding a constant to the phase of all wave functions, everywhere); this is a global symmetry. In quantum electrodynamics, the theory is also invariant under a *local* change of phase, that is – one may shift the phase of all wave functions so that the shift may be different at every point in space-time. This is a *local* symmetry. However, in order for a well-defined derivative operator to exist, one must introduce a new field, the gauge field, which also transforms in order for the local change of variables (the phase in our example) not to affect the derivative. In quantum electrodynamics, this gauge field is the electromagnetic field. The change of local gauge of variables is termed gauge transformation.

By Noether's theorem, for every such symmetry there exists an associated conserved current. The aforementioned symmetry of the wavefunction under global phase changes implies the conservation of electric charge. Since the excitations of fields represent particles, the particle associated with excitations of the gauge field is the gauge boson, e.g., the photon in the case of quantum electrodynamics.

The degrees of freedom in quantum field theory are local fluctuations of the fields. The existence of a gauge symmetry reduces the number of degrees of freedom, simply because some fluctuations of the fields can be transformed to zero by gauge transformations, so they are equivalent to having no fluctuations at all, and they, therefore, have no physical meaning. Such fluctuations are usually called "non-physical degrees of freedom" or *gauge artifacts*; usually, some of them have a negative norm, making them inadequate for a consistent theory. Therefore, if a classical field theory has a gauge symmetry, then its quantized version (the corresponding quantum field theory) will have this symmetry as well. In other words, a gauge symmetry cannot have a quantum anomaly.

In general, the gauge transformations of a theory consist of several different transformations, which may not be commutative. These transformations are combine into the framework of a gauge group; infinitesimal gauge transformations are the gauge group generators. Thus, the number of gauge bosons is the group dimension (i.e., the number of generators forming the basis of the corresponding Lie algebra).

All the known fundamental interactions in nature are described by gauge theories (possibly barring the Higgs multiplet couplings, if considered in isolation). These are:

- Quantum chromodynamics, whose gauge group is $SU(3)$. The gauge bosons are eight gluons.
- The electroweak theory, whose gauge group is $U(1) \times SU(2)$, (a direct product of $U(1)$ and $SU(2)$). The gauge bosons are the photon and the massive $W^{\pm}$ and Z^{o} bosons.
- Gravity, whose classical theory is general relativity, relies on the equivalence principle, which is essentially a form of gauge symmetry. Its action may also be written as a gauge theory of the Lorentz group on tangent space.

Supersymmetry

Supersymmetry assumes that every fundamental fermion has a superpartner that is a boson and vice versa. Its gauge theory, Supergravity, is an extension of general relativity. Supersymmetry is a key ingredient for the consistency of string theory.

It was utilized in order to solve the so-called Hierarchy Problem of the standard model, that is, to explain why particles not protected by any symmetry (like the Higgs boson) do not receive radiative corrections to their mass, driving it to the larger scales such as that of GUTs, or the Planck mass of gravity. The way supersymmetry protects scale hierarchies is the following: since for every particle there is a superpartner with the same mass but different statistics, any loop in a radiative correction is cancelled by the loop corresponding to its superpartner, rendering the theory more UV finite.

Since, however, no super partners have been observed, if supersymmetry existed it should be broken severely (through a so-called soft term, which breaks supersymmetry without ruining its helpful features). The simplest models of this breaking require that the energy of the superpartners not be too high; in these cases, supersymmetry could be observed by experiments at the Large Hadron Collider. However, to date, after the observation of the Higgs boson there, no such superparticles have been discovered.

Axiomatic Approaches

The preceding description of quantum field theory follows the spirit in which most physicists approach the subject. However, it is not mathematically rigorous. Over the past several decades, there have been many attempts to put quantum field theory on a firm mathematical footing by formulating a set of axioms for it. Finding proper axioms for quantum field theory is still an open and difficult problem in mathematics. One of the Millennium Prize Problems—proving the existence of a mass gap in Yang–Mills theory—is linked to this issue. These attempts fall into two broad classes.

Wightman Axioms

The first class of axioms, first proposed during the 1950s, include the Wightman, Osterwalder–Schrader, and Haag–Kastler systems. They attempted to formalize the physicists' notion of an "operator-valued field" within the context of functional analysis and enjoyed limited success. It was possible to prove that any quantum field theory satisfying these axioms satisfied certain general theorems, such as the spin-statistics theorem and the CPT theorem. Unfortunately, it proved extraordinarily difficult to show that any realistic field theory, including the Standard Model, satisfied these axioms. Most of the theories that could be treated with these analytic axioms were physically trivial, being restricted to low-dimensions and lacking interesting dynamics. The construction of theories satisfying one of these sets of axioms falls in the field of constructive quantum field theory. Important work was done in this area in the 1970s by Segal, Glimm, Jaffe and others.

Topological Quantum Field Theory

During the 1980s, the second set of axioms based on topological ideas was proposed. Before 1980 all states of matter could be classified by geometry and the principle of broken symmetry. For example Einstein's theory of general relativity is based on the geometrical curvature of space and time, while crystals, magnets and superconductors can all be classified by the symmetries they break. In 1980 the quantum Hall effect provided the first example of a state of matter that has no spontaneous broken symmetry; its characterization is dependant on its topology and not on its geometry. The quantum Hall effect can be described by extending quantum field theory into an effective topological quantum field theory based on the Chern–Simons theory. This line of investigation, which extends quantum field theories to topological quantum field theories, is associated most closely with Michael Atiyah and Graeme Segal, and was notably expanded upon by Edward Witten, Richard Borcherds, and Maxim Kontsevich. The main impact of topological quantum field theory has been in condensed matter physics where physicists have observed exotic quasiparticles such as magnetic monopoles and Majorana fermions. Topological field considerations could have radical applications in a new form of electronics called spintronics and topological quantum computers. The Standard Model allows for topological terms but is generally not formulated as a topological quantum field theory. Topological quantum field theory has also had broad impact in mathematics, with important applications in representation theory, algebraic topology, and differential geometry.

Haag's Theorem

From a mathematically rigorous perspective, there exists no interaction picture in a Lorentz-covariant quantum field theory. This implies that the perturbative approach of Feynman diagrams in QFT is not strictly justified, despite producing vastly precise predictions validated by experiment. This is called Haag's theorem, but most particle physicists relying on QFT largely shrug it off, as not really limiting the power of the theory.

Ginzburg–Landau Theory

In physics, Ginzburg–Landau theory, often called Landau–Ginzburg theory, named after Vitaly Lazarevich Ginzburg and Lev Landau, is a mathematical physical theory used to describe superconductivity. In its initial form, it was postulated as a phenomenological model which could describe type-I superconductors without examining their microscopic properties. Later, a version of Ginzburg–Landau theory was derived from the Bardeen–Cooper–Schrieffer microscopic theory by Lev Gor'kov, thus showing that it also appears in some limit of microscopic theory and giving microscopic interpretation of all its parameters.

Introduction

Based on Landau's previously-established theory of second-order phase transitions, Ginzburg and Landau argued that the free energy, F, of a superconductor near the superconducting transition can be expressed in terms of a complex order parameter field, ψ, which is nonzero below a phase transition into a superconducting state and is related to the density of the superconducting component, although no direct interpretation of this parameter was given in the original paper. Assuming smallness of $|\psi|$ and smallness of its gradients, the free energy has the form of a field theory.

$$F = F_n + \alpha|\psi|^2 + \frac{\beta}{2}|\psi|^4 + \frac{1}{2m}\left|\left(-i\hbar\nabla - 2e\mathbf{A}\right)\psi\right|^2 + \frac{|\mathbf{B}|^2}{2\mu_0}$$

where Fn is the free energy in the normal phase, α and β in the initial argument were treated as phenomenological parameters, m is an effective mass, e is the charge of an electron, A is the magnetic vector potential, and $\mathbf{B} = \nabla\times\mathbf{A}$ is the magnetic field. By minimizing the free energy with respect to variations in the order parameter and the vector potential, one arrives at the Ginzburg–Landau equations

$$\alpha\psi + \beta|\psi|^2\psi + \frac{1}{2m}\left(-i\hbar\nabla - 2e\mathbf{A}\right)^2\psi = 0$$

$$\nabla \times \mathbf{B} = \mu_0 \mathbf{j} \ ; \ \mathbf{j} = \frac{2e}{m} \mathrm{Re}\{\psi^* (-i\hbar\nabla - 2e\mathbf{A})\psi\}$$

where $\mathbf{j}$ denotes the dissipation-less electric current density and *Re* the *real part*. The first equation — which bears some similarities to the time-independent Schrödinger equation, but is principally different due to a nonlinear term — determines the order parameter, ψ. The second equation then provides the superconducting current.

Simple Interpretation

Consider a homogeneous superconductor where there is no superconducting current and the equation for ψ simplifies to:

$$\alpha\psi + \beta|\psi|^2\psi = 0.$$

This equation has a trivial solution: $\psi = 0$. This corresponds to the normal state of the superconductor, that is for temperatures above the superconducting transition temperature, $T > T_c$.

Below the superconducting transition temperature, the above equation is expected to have a non-trivial solution (that is $\psi \neq 0$). Under this assumption the equation above can be rearranged into:

$$|\psi|^2 = -\frac{\alpha}{\beta}.$$

When the right hand side of this equation is positive, there is a nonzero solution for ψ (remember that the magnitude of a complex number can be positive or zero). This can be achieved by assuming the following temperature dependence of α: $\alpha(T) = \alpha_0 (T - T_c)$ with $\alpha_0/\beta > 0$:

- Above the superconducting transition temperature, $T > T_c$, the expression $\alpha(T)/\beta$ is positive and the right hand side of the equation above is negative. The magnitude of a complex number must be a non-negative number, so only $\psi = 0$ solves the Ginzburg–Landau equation.

- Below the superconducting transition temperature, $T < T_c$, the right hand side of the equation above is positive and there is a non-trivial solution for ψ. Furthermore,

  $$|\psi|^2 = -\frac{\alpha_0(T - T_c)}{\beta},$$

  that is ψ approaches zero as T gets closer to T_c from below. Such a behaviour is typical for a second order phase transition.

In Ginzburg–Landau theory the electrons that contribute to superconductivity were proposed to form a superfluid. In this interpretation, $|\psi|^2$ indicates the fraction of electrons that have condensed into a superfluid.

Coherence Length and Penetration Depth

The Ginzburg–Landau equations predicted two new characteristic lengths in a superconductor which was termed coherence length, ξ. For $T > T_c$ (normal phase), it is given by

$$\xi = \sqrt{\frac{\hbar^2}{2m|\alpha|}}.$$

while for $T < T_c$ (superconducting phase), where it is more relevant, it is given by

$$\xi = \sqrt{\frac{\hbar^2}{4m|\alpha|}}.$$

It sets the exponential law according to which small perturbations of density of superconducting electrons recover their equilibrium value ψ_0. Thus this theory characterized all superconductors by two length scales. The second one is the penetration depth, λ. It was previously introduced by the London brothers in their London theory. Expressed in terms of the parameters of Ginzburg–Landau model it is

$$\lambda = \sqrt{\frac{m}{4\mu_0 e^2 \psi_0^2}},$$

where ψ_0 is the equilibrium value of the order parameter in the absence of an electromagnetic field. The penetration depth sets the exponential law according to which an external magnetic field decays inside the superconductor.

The original idea on the parameter "k" belongs to Landau. The ratio $\kappa = \lambda/\xi$ is presently known as the Ginzburg–Landau parameter. It has been proposed by Landau that Type I superconductors are those with $0 < \kappa < 1/\sqrt{2}$, and Type II superconductors those with $\kappa > 1/\sqrt{2}$.

The exponential decay of the magnetic field is equivalent with the Higgs mechanism in high-energy physics.

Fluctuations in the Ginzburg–Landau Model

Taking into account fluctuations. For Type II superconductors, the phase transition from the normal state is of second order, as demonstrated by Dasgupta and Halperin. While for Type I superconductors it is of first order as demonstrated by Halperin, Lubensky and Ma.

Classification of Superconductors based on Ginzburg–Landau Theory

In the original paper Ginzburg and Landau observed the existence of two types of superconductors depending on the energy of the interface between the normal and superconducting states. The Meissner state breaks down when the applied magnetic field is too large. Superconductors can be divided into two classes according to how this breakdown occurs. In Type I superconductors, superconductivity is abruptly destroyed when the strength of the applied field rises above a critical value H_c. Depending on the geometry of the sample, one may obtain an intermediate state consisting of a baroque pattern of regions of normal material carrying a magnetic field mixed with regions of superconducting material containing no field. In Type II superconductors, raising the applied field past a critical value H_{c1} leads to a mixed state (also known as the vortex state) in which an increasing amount of magnetic flux penetrates the material, but there remains no resistance to the flow of electric current as long as the current is not too large. At a second critical field strength H_{c2}, superconductivity is destroyed. The mixed state is actually caused by vortices in the electronic superfluid, sometimes called fluxons because the flux carried by these vortices is quantized. Most pure elemental superconductors, except niobium and carbon nanotubes, are Type I, while almost all impure and compound superconductors are Type II.

The most important finding from Ginzburg–Landau theory was made by Alexei Abrikosov in 1957. He used Ginzburg–Landau theory to explain experiments on superconducting alloys and thin films. He found that in a type-II superconductor in a high magnetic field, the field penetrates in a triangular lattice of quantized tubes of flux vortices.

Landau–Ginzburg Theories in String Theory

In particle physics, any quantum field theory with a unique classical vacuum state and a potential energy with a degenerate critical point is called a Landau–Ginzburg theory. The generalization to $N = (2,2)$ supersymmetric theories in 2 spacetime dimensions was proposed by Cumrun Vafa and Nicholas Warner in the November 1988 article Catastrophes and the Classification of Conformal Theories, in this generalization one imposes that the superpotential possess a degenerate critical point. The same month, together with Brian Greene they argued that these theories are related by a renormalization group flow to sigma models on Calabi–Yau manifolds in the paper Calabi–Yau Manifolds and Renormalization Group Flows. In his 1993 paper Phases of $N = 2$ theories in two-dimensions, Edward Witten argued that Landau–Ginzburg theories and sigma models on Calabi–Yau manifolds are different phases of the same theory. A construction of such a duality was given by relating the Gromov–Witten theory of Calabi–Yau orbifolds to FJRW theory an analogous Landau–Ginzburg "FJRW" theory in The Witten Equation, Mirror Symmetry and Quantum Singularity Theory. Witten's sigma models were later used to describe the low energy dynamics of 4-dimensional gauge theories with monopoles as well as brane constructions. Gaiotto, Gukov & Seiberg (2013)

Bloch Wave

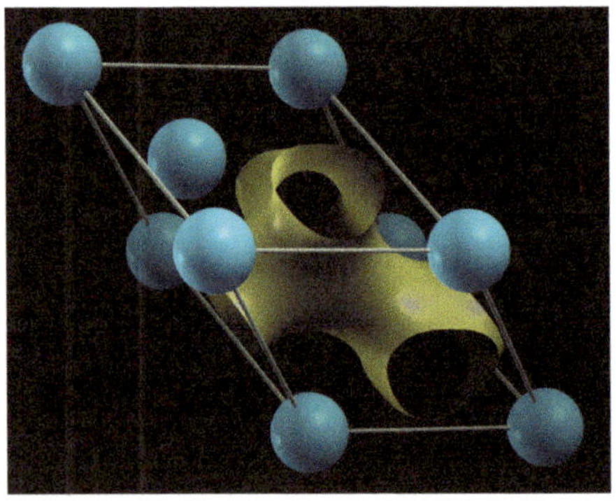

Isosurface of the square modulus of a Bloch wave in silicon lattice

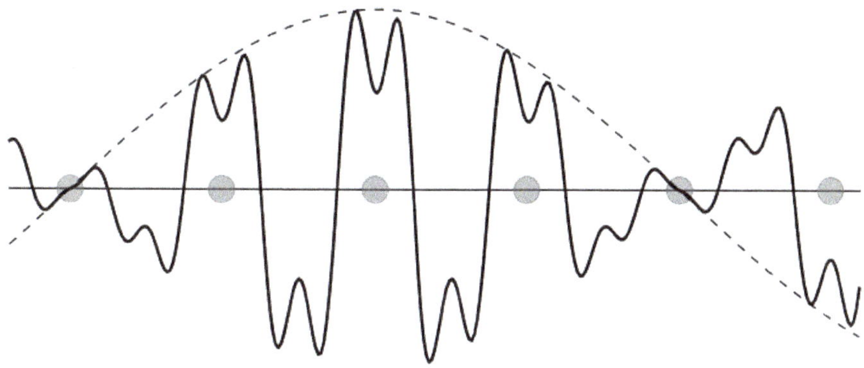

Solid line: A schematic of a typical Bloch wave in one dimension. (The actual wave is complex; this is the real part.) The dotted line is from the $e^{ik\cdot r}$ factor. The light circles represent atoms.

A Bloch wave (also called Bloch state or Bloch function or Bloch wave function), named after Swiss physicist Felix Bloch, is a type of wavefunction for a particle in a periodically-repeating environment, most commonly an electron in a crystal. A wavefunction ψ is a Bloch wave if it has the form:

$$\psi(\mathbf{r}) = e^{i\mathbf{k}\cdot\mathbf{r}}u(\mathbf{r})$$

where r is position, ψ is the Bloch wave, *u* is a periodic function with the same periodicity as the crystal, k is a vector of real numbers called the crystal wave vector, *e* is Euler's number, and *i* is the imaginary unit. In other words, if we multiply a plane wave by a periodic function, we get a Bloch wave.

Bloch waves are important because of Bloch's theorem, which states that the energy eigenstates for an electron in a crystal can be written as Bloch waves. (More precisely, it states that the electron wave functions in a crystal have a basis consisting entirely of Bloch wave energy eigenstates.) This fact underlies the concept of electronic band structures.

These Bloch wave energy eigenstates are written with subscripts as $\psi_{n\,k}$, where *n* is a discrete index, called the *band index*, which is present because there are many different Bloch waves with the same k (each has a different periodic component *u*). Within a band (i.e., for fixed *n*), $\psi_{n\,k}$ varies continuously with k, as does its energy. Also, for any reciprocal lattice vector K, $\psi_{n\,k} = \psi_{n,(k+K)}$. Therefore, all distinct Bloch waves occur for k-values within the first Brillouin zone of the reciprocal lattice.

Applications and Consequences

Applicability

The most common example of Bloch's theorem is describing electrons in a crystal. However, a Bloch-wave description applies more generally to any wave-like phenomenon in a periodic medium. For example, a periodic dielectric in electromagnetism leads to photonic crystals, and a periodic acoustic medium leads to phononic crystals. It is generally treated in the various forms of the dynamical theory of diffraction.

Meaning and Non-uniqueness of the k-vector

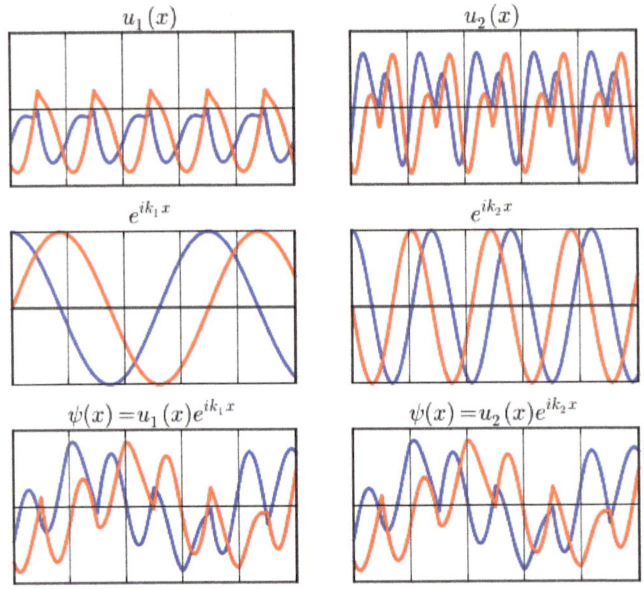

A Bloch wave (bottom) can be broken up into the product of a periodic function (top) and a plane-wave (center). The left side and right side represent the same Bloch wave broken up in two different ways, involving the wave vector k_1 (left) or k_2 (right). The difference $(k_1 - k_2)$ is a reciprocal lattice vector. In all plots, blue is real part and red is imaginary part.

Suppose an electron is in a Bloch state

$$\psi(\mathbf{r}) = e^{i\mathbf{k}\cdot\mathbf{r}} u(\mathbf{r}),$$

where u is periodic with the same periodicity as the crystal lattice. The actual quantum state of the electron is entirely determined by ψ, not k or u directly. This is important because k and u are *not* unique. Specifically, if ψ can be written as above using k, it can *also* be written using (k + K), where K is any reciprocal lattice vector (see figure at right). Therefore, wave vectors that differ by a reciprocal lattice vector are equivalent, in the sense that they characterize the same set of Bloch states.

The first Brillouin zone is a restricted set of k-vectors with the property that no two of them are equivalent, yet every possible k is equivalent to one (and only one) vector in the first Brillouin zone. Therefore, if we restrict k to the first Brillouin zone, then every Bloch state has a unique k. Therefore, the first Brillouin zone is often used to depict all of the Bloch states without redundancy, for example in a band structure, and it is used for the same reason in many calculations.

When k is multiplied by the reduced Planck's constant, it equals the electron's crystal momentum. Related to this, the group velocity of an electron can be calculated based on how the energy of a Bloch state varies with k.

Proof of Bloch's Theorem

Next, we prove Bloch's theorem:

For electrons in a perfect crystal, there is a basis of wavefunctions with the properties:

- Each of these wavefunctions is an energy eigenstate
- Each of these wavefunctions is a Bloch wave, meaning that this wavefunction ψ can be written in the form

$$(\mathbf{r}) \quad e^{\mathbf{k}\cdot\mathbf{r}}u(\mathbf{r})$$

where u has the same periodicity as the atomic structure of the crystal.

Preliminaries: Crystal Symmetries, Lattice, and Reciprocal Lattice

The defining property of a crystal is translational symmetry, which means that if the crystal is shifted an appropriate amount, it winds up with all its atoms in the same places. (A finite-size crystal cannot have perfect translational symmetry, but it is a useful approximation).

A three-dimensional crystal has three *primitive lattice vectors* a_1, a_2, a_3. If the crystal is shifted by any of these three vectors, or a combination of them of the form

$$n_1\mathbf{a}_1 + n_2\mathbf{a}_2 + n_3\mathbf{a}_3$$

where n_i are three integers, then the atoms end up in the same set of locations as they started.

Another helpful ingredient in the proof is the *reciprocal lattice vectors*. These are three vectors b_1, b_2, b_3 (with units of inverse length), with the property that $a_i \cdot b_i = 2\pi$, but $a_i \cdot b_j = 0$ when $i \neq j$.

Lemma about Translation Operators

Let $\hat{T}_{n_1,n_2,n_3}$ denote a translation operator that shifts every wave function by the amount n1a1 + n2a2 + n3a3 (as above, nj are integers). The following fact is helpful for the proof of Bloch's theorem:

> Lemma: If a wavefunction ψ is an eigenstate of all of the translation operators (simultaneously), then ψ is a Bloch wave.

Proof: Assume that we have a wavefunction ψ which is an eigenstate of all the translation operators. As a special case of this,

$$\psi(\mathbf{r} + \mathbf{a}_j) = C_j\psi(\mathbf{r})$$

for $j = 1, 2, 3$, where C_j are three numbers (the eigenvalues) which do not depend on **r**. It is helpful to write the numbers C_j in a different form, by choosing three numbers θ_1, θ_2, θ_3 with $e^{2\pi i\theta_j} = C_j$:

$$\psi(\mathbf{r} + \mathbf{a}_j) = e^{2\pi i\theta_j}\psi(\mathbf{r})$$

Again, the θ_j are three numbers which do not depend on r. Define $k = \theta_1 b_1 + \theta_2 b_2 + \theta_3 b_3$, where b_j are the reciprocal lattice vectors. Finally, define

$$u(\mathbf{r}) = e^{-i\mathbf{k}\cdot\mathbf{r}}\psi(\mathbf{r}).$$

Then

$$u(\mathbf{r}+\mathbf{a}_j)=e^{-i\mathbf{k}\cdot(\mathbf{r}+\mathbf{a}_j)}\psi(\mathbf{r}+\mathbf{a}_j)=\left(e^{-i\mathbf{k}\cdot\mathbf{r}}e^{-i\mathbf{k}\cdot\mathbf{a}_j}\right)\left(e^{2\pi i\theta_j}\psi(\mathbf{r})\right)=e^{-i\mathbf{k}\cdot\mathbf{r}}e^{-2\pi i\theta_j}e^{2\pi i\theta_j}\psi(\mathbf{r})=u(\mathbf{r}).$$

This proves that u has the periodicity of the lattice. Since $\psi(\mathrm{r}) = e^{i\mathrm{k}\cdot\mathrm{r}}u(\mathrm{r})$, that proves that the state is a Bloch wave.

Proof

Finally, we are ready for the main proof of Bloch's theorem.

As above, let $\hat{T}_{n_1,n_2,n_3}$ denote a translation operator that shifts every wave function by the amount n1a1 + n2a2 + n3a3, where ni are integers. Because the crystal has translational symmetry, this operator commutes with the Hamiltonian operator. Moreover, every such translation operator commutes with every other. Therefore, there is a simultaneous eigenbasis of the Hamiltonian operator and every possible $\hat{T}_{n_1,n_2,n_3}$ operator. This basis is what we are looking for. The wavefunctions in this basis are energy eigenstates (because they are eigenstates of the Hamiltonian), and they are also Bloch waves (because they are eigenstates of the translation operators).

History and Related Equations

The concept of the Bloch state was developed by Felix Bloch in 1928, to describe the conduction of electrons in crystalline solids. The same underlying mathematics, however, was also discovered independently several times: by George William Hill (1877), Gaston Floquet (1883), and Alexander Lyapunov (1892). As a result, a variety of nomenclatures are common: applied to ordinary differential equations, it is called Floquet theory (or occasionally the *Lyapunov–Floquet theorem*). The general form of a one-dimensional periodic potential equation is Hill's equation:

$$\frac{d^2y}{dt^2}+f(t)y=0,$$

where *f(t)* is a periodic potential. Specific periodic one-dimensional equations include the Kronig–Penney model and Mathieu's equation.

Mathematically Bloch's theorem is interpreted in terms of unitary characters of a lattice group, and is applied to spectral geometry.

Fermi Gas

A Fermi gas is a phase of matter which is an ensemble of a large number of fermions (named after Enrico Fermi). Fermions are particles that obey Fermi–Dirac statistics. These statistics determine the energy distribution of fermions in a Fermi gas in thermal equilibrium, and is characterized by their number density, temperature, and the set of available energy states.

By the Pauli exclusion principle, no quantum state can be occupied by more than one fermion with an identical set of quantum numbers. Thus a non-interacting Fermi gas, unlike a Bose gas, is prohibited from condensing into a Bose–Einstein condensate, although interacting Fermi gases might. The total energy of the Fermi gas at absolute zero is larger than the sum of the single-particle ground states because the Pauli principle implies a sort of interaction or pressure that keeps fermions separated and moving. For this reason, the pressure of a Fermi gas is non-zero even at zero temperature, in contrast to that of a classical ideal gas. This so-called degeneracy pressure stabilizes a neutron star (a Fermi gas of neutrons) or a white dwarf star (a Fermi gas of electrons) against the inward pull of gravity, which would ostensibly collapse the star into a black hole. Only when a star is sufficiently massive to overcome the degeneracy pressure can it collapse into a singularity.

It is possible to define a Fermi temperature below which the gas can be considered degenerate (its pressure derives almost exclusively from the Pauli principle). This temperature depends on the mass of the fermions and the density of energy states. For metals, the electron gas's Fermi temperature is generally many thousands of kelvins, so in human applications they can be considered degenerate. The maximum energy of the fermions at zero temperature is called the Fermi energy. The Fermi energy surface in momentum space is known as the Fermi surface.

Ideal Fermi Gas

An ideal Fermi gas or free Fermi gas is a physical model assuming a collection of non-interacting fermions. It is the quantum mechanical version of an ideal gas, for the case of fermionic particles. The behavior of electrons in a white dwarf or neutrons in a neutron star can be approximated by treating them as an ideal Fermi gas. Something similar can be done for periodic systems, such as electrons moving in the crystal lattice of metals and semiconductors, using the so-called *quasi-momentum* or *crystal momentum* (Bloch wave). Since interactions are neglected by definition, the problem of treating the equilibrium properties and dynamics of an ideal Fermi gas reduces to the study of the behavior of single independent particles. As such, it is still relatively tractable and forms the starting point for more advanced theories that deal with interactions, e.g., using the perturbation theory.

Assuming that the concentration of fermions does not change with temperature, then the total chemical potential μ (Fermi level) of the three-dimensional ideal Fermi gas is related to the zero temperature Fermi energy EF by a Sommerfeld expansion (assuming $kT \ll E_F$):

$$\mu = E_0 + E_F\left[1 - \frac{\pi^2}{12}\left(\frac{kT}{E_F}\right)^2 - \frac{\pi^4}{80}\left(\frac{kT}{E_F}\right)^4 + \cdots\right]$$

where E_0 is the potential energy per particle, k is the Boltzmann constant and T is temperature.

Hence, the internal chemical potential, μ-E_0, is approximately equal to the Fermi energy at temperatures that are much lower than the characteristic Fermi temperature E_F/k. The characteristic temperature is on the order of 10^5 K for a metal, hence at room temperature (300 K), the Fermi energy and internal chemical potential are essentially equivalent.

Fermi Liquid Theory

Fermi liquid theory (also known as Landau–Fermi liquid theory) is a theoretical model of interacting fermions that describes the normal state of most metals at sufficiently low temperatures. The interaction between the particles of the many-body system does not need to be small. The phenomenological theory of Fermi liquids was introduced by the Soviet physicist Lev Davidovich Landau in 1956, and later developed by Alexei Abrikosov and Isaak Khalatnikov using diagrammatic perturbation theory. The theory explains why some of the properties of an interacting fermion system are very similar to those of the Fermi gas (i.e. non-interacting fermions), and why other properties differ.

Important examples of where Fermi liquid theory has been successfully applied are most notably electrons in most metals and Liquid He-3. Liquid He-3 is a Fermi liquid at low temperatures (but not low enough to be in its superfluid phase). He-3 is an isotope of helium, with 2 protons, 1 neutron and 2 electrons per atom. Because there is an odd number of fermions inside the nucleus, the atom itself is also a fermion. The electrons in a normal (non-superconducting) metal also form a Fermi liquid, as do the nucleons (protons and neutrons) in an atomic nucleus. Strontium ruthenate displays some key properties of Fermi liquids, despite being a strongly correlated material, and is compared with high temperature superconductors like cuprates.

Description

The key ideas behind Landau's theory are the notion of *adiabaticity* and the exclusion principle. Consider a non-interacting fermion system (a Fermi gas), and suppose we "turn on" the interaction slowly. Landau argued that in this situation, the ground state of the Fermi gas would adiabatically transform into the ground state of the interacting system.

By Pauli's exclusion principle, the ground state Ψ_0 of a Fermi gas consists of fermions occupying all momentum states corresponding to momentum $p < p_F$ with all higher momentum states unoccupied. As interaction is turned on, the spin, charge and momentum of the fermions corresponding to the occupied states remain unchanged, while their dynamical properties, such as their mass, magnetic moment etc. are renormalized to new values. Thus, there is a one-to-one correspondence between the elementary excitations of a Fermi gas system and a Fermi liquid system. In the context of Fermi liquids, these excitations are called "quasi-particles".

Landau quasiparticles are long-lived excitations with a lifetime τ that satisfies $\frac{\hbar}{\tau} \ll \epsilon_p$ where ϵ_p is the quasiparticle energy (measured from the Fermi energy). At finite temperature, ϵ_p is on the order of the thermal energy $k_B T$, and the condition for Landau quasiparticles can be reformulated as $\frac{\hbar}{\tau} \ll k_B T$.

For this system, the Green's function can be written (near its poles) in the form

$$G(\omega, p) \approx \frac{Z}{\omega + \mu - \epsilon(p)}$$

where μ is the chemical potential and $\epsilon(p)$ is the energy corresponding to the given momentum state.

The value Z is called the quasiparticle residue and is very characteristic of Fermi liquid theory. The spectral function for the system can be directly observed via ARPES experiment, and can be written (in the limit of low-lying excitations) in the form:

$$A(\vec{k},\omega) = Z\delta(\omega - v_F k_{\|})$$

where v_F is the Fermi velocity.

Physically, we can say that a propagating fermion interacts with its surrounding in such a way that the net effect of the interactions is to make the fermion behave as a "dressed" fermion, altering its effective mass and other dynamical properties. These "dressed" fermions are what we think of as "quasiparticles".

Another important property of Fermi liquids is related to the scattering cross section for electrons. Suppose we have an electron with energy ϵ_1 above the Fermi surface, and suppose it scatters with a particle in the Fermi sea with energy ϵ_2. By Pauli's exclusion principle, both the particles after scattering have to lie above the Fermi surface, with energies $\epsilon_3, \epsilon_4 > \epsilon_F$ Now, suppose the initial electron has energy very close to the Fermi surface $\epsilon \approx \epsilon_F$ Then, we have that $\epsilon_2, \epsilon_3, \epsilon_4$ also have to be very close to the Fermi surface. This reduces the phase space volume of the possible states after scattering, and hence, by Fermi's golden rule, the scattering cross section goes to zero. Thus we can say that the lifetime of particles at the Fermi surface goes to infinity.

Similarities to Fermi Gas

The Fermi liquid is qualitatively analogous to the non-interacting Fermi gas, in the following sense: The system's dynamics and thermodynamics at low excitation energies and temperatures may be described by substituting the non-interacting fermions with interacting quasiparticles, each of which carries the same spin, charge and momentum as the original particles. Physically these may be thought of as being particles whose motion is disturbed by the surrounding particles and which themselves perturb the particles in their vicinity. Each many-particle excited state of the interacting system may be described by listing all occupied momentum states, just as in the non-interacting system. As a consequence, quantities such as the heat capacity of the Fermi liquid behave qualitatively in the same way as in the Fermi gas (e.g. the heat capacity rises linearly with temperature).

Differences From Fermi Gas

The following differences to the non-interacting Fermi gas arise:

Energy

The energy of a many-particle state is not simply a sum of the single-particle energies of all occupied states. Instead, the change in energy for a given change δn_k in occupation of states k contains terms both linear and quadratic in δn_k (for the Fermi gas, it would only be linear, $\delta n_k \epsilon_k$, where ϵ_k denotes the single-particle energies). The linear contribution corresponds to renormalized single-particle energies, which involve, e.g., a change in the effective mass of particles. The quadratic terms correspond to a sort of "mean-field" interaction between quasiparticles, which is parameterized by so-called Landau Fermi liquid parameters and determines the behaviour of density oscillations (and spin-density oscillations) in the Fermi liquid. Still, these mean-field interactions do not lead to a scattering of quasi-particles with a transfer of particles between different momentum states.

The renormalization of the mass of a fluid of interacting fermions can be calculated from first principles using many-body computational techniques. For the two-dimensional homogeneous electron gas, *GW* calculations and quantum Monte Carlo methods have been used to calculate renormalized quasiparticle effective masses.

Specific Heat and Compressibility

Specific heat, compressibility, spin-susceptibility and other quantities show the same qualitative behaviour (e.g. dependence on temperature) as in the Fermi gas, but the magnitude is (sometimes strongly) changed.

Interactions

In addition to the mean-field interactions, some weak interactions between quasiparticles remain, which lead to scattering of quasiparticles off each other. Therefore, quasiparticles acquire a finite lifetime. However, at low enough energies above the Fermi surface, this lifetime becomes very long, such that the product of excitation energy (expressed in frequency) and lifetime is much larger than one. In this sense, the quasiparticle energy is still well-defined (in the opposite limit, Heisenberg's uncertainty relation would prevent an accurate definition of the energy).

Structure

The structure of the "bare" particle's (as opposed to quasiparticle) Green's function is similar to that in the Fermi gas (where, for a given momentum, the Green's function in frequency space is a delta peak at the respective single-particle energy). The delta peak in the density-of-states is broadened (with a width given by the quasiparticle lifetime). In addition (and in contrast to the quasiparticle Green's function), its weight (integral over frequency) is suppressed by a quasiparticle weight factor $0 < Z < 1$. The remainder of the total weight is in a broad "incoherent background", corresponding to the strong effects of interactions on the fermions at short time-scales.

Distribution

The distribution of particles (as opposed to quasiparticles) over momentum states at zero temperature still shows a discontinuous jump at the Fermi surface (as in the Fermi gas), but it does not drop from 1 to 0: the step is only of size Z.

Electrical Resistivity

In a metal the resistivity at low temperatures is dominated by electron-electron scattering in combination with umklapp scattering. For a Fermi liquid, the resistivity from this mechanism varies as T^2, which is often taken as an experimental check for Fermi liquid behaviour (in addition to the linear temperature-dependence of the specific heat), although it only arises in combination with the lattice. In certain cases, umklapp scattering is not required. For example, the resistivity of compensated semimetals scales as T^2 because of mutual scattering of electron and hole. This is known as the Baber mechanism.

Optical Response

Fermi liquid theory predicts that the scattering rate, which governs the optical response of metals, not only depends quadratically on temperature (thus causing the T^2 dependence of the dc resistance), but it also depends quadratically on frequency. This is in contrast to the Drude prediction for non-interacting metallic electrons, where the scattering rate is a constant as a function of frequency. One material in which optical Fermi liquid behavior was experimentally observed is the low-temperature metallic phase of Sr_2RuO_4.

Instabilities of the Fermi Liquid

The experimental observation of exotic phases in strongly correlated systems has triggered an enormous effort from the theoretical community to try to understand their microscopic origin. One possible route to detect instabilities of a Fermi liquid is precisely the analysis done by Pomeranchuk. Due to that, the Pomeranchuk instability has been studied by several authors with different techniques in the last few years and in particular, the instability of the Fermi liquid towards the nematic phase was investigated for several models.

Non-Fermi Liquids

The term non-Fermi liquid, also known as "strange metal", is used to describe a system which displays breakdown of Fermi-liquid behaviour. The simplest example of such a system is the system of interacting fermions in one-dimension, called *Luttinger liquid*. Although Luttinger liquids are physically similar to Fermi liquids, the restriction to one dimension gives rise to several qualitative differences such as the absence of a *quasiparticle peak* in the momentum dependent spectral function, spin-charge separation, and the presence of spin density waves. One cannot ignore the existence of interactions in one-dimension and has to describe the problem with a non-Fermi theory, where Luttinger liquid is one of them. At small finite spin-temperatures in one-dimension the ground-state of the system is described by spin-incoherent Luttinger liquid (SILL).

Another example of such behaviour is observed at quantum critical points of certain second-order phase transitions, such as Heavy fermion criticality, Mott criticality and high-T_c cuprate phase transitions. The ground state of such transitions is characterized by the presence of a sharp Fermi surface, although there may not be well-defined quasiparticles. That is, on approaching the critical point, it is observed that the quasiparticle residue $Z \to 0$

Understanding the behaviour of non-Fermi liquids is an important problem in condensed matter physics. Approaches towards explaining these phenomena include the treatment of *marginal Fermi liquids*; attempts to understand critical points and derive scaling relations; and descriptions using *emergent* gauge theories with techniques of holographic gauge/gravity duality.

Many-body Theory

The theory of Quantum Thermodynamic Motion (or QTM) is an area of physics which provides a collective framework of understanding for the behavior of complex assemblies, namely their constitute particles and force interactions. In general terms, the many-body theory describes effects that manifest themselves in a system which contains a large numbers of non-trivial forces (e.g. particles and fields). While the elementary laws of physics that govern the bodies of motion on each individual particle may or may not be trivial, the study of systems collective particles may display extremely complex phenomena. As often is the case in which a complex array of forces display emergent phenomenon which oft bear little or no resemblance to the underlying system dynamics.

BCS Theory

BCS theory or Bardeen–Cooper–Schrieffer theory (named after John Bardeen, Leon Cooper, and John Robert Schrieffer) is the first microscopic theory of superconductivity since Heike Kamerlingh Onnes's 1911 discovery. The theory describes superconductivity as a microscopic effect caused by a condensation of Cooper pairs into a boson-like state. The theory is also used in nuclear physics to describe the pairing interaction between nucleons in an atomic nucleus.

It was proposed by Bardeen, Cooper, and Schrieffer in 1957; they received the Nobel Prize in Physics for this theory in 1972.

History

Rapid progress in understanding superconductivity gained momentum in the mid-1950s. It began with the 1948 paper, "On the Problem of the Molecular Theory of Superconductivity", where Fritz London proposed that the phenomenological London equations may be consequences of the coherence of a quantum state. In 1953, Brian Pippard, motivated by penetration experiments, proposed that this would modify the London equations via a new scale parameter called the coherence length. John Bardeen then argued in the 1955 paper, "Theory of the Meissner Effect in Superconductors", that such a modification naturally occurs in a theory with an energy gap. The key ingredient was Leon Neil Cooper's calculation of the bound states of electrons subject to an attractive force in his 1956 paper, "Bound Electron Pairs in a Degenerate Fermi Gas".

In 1957 Bardeen and Cooper assembled these ingredients and constructed such a theory, the BCS theory, with Robert Schrieffer. The theory was first published in April 1957 in the letter, "Microscopic theory of superconductivity". The demonstration that the phase transition is second order, that it reproduces the Meissner effect and the calculations of specific heats and penetration depths appeared in the December 1957 article, "Theory of superconductivity". They received the Nobel Prize in Physics in 1972 for this theory.

In 1986, high-temperature superconductivity was discovered in some materials at temperatures up to about 130 K, considerably above the previous limit of about 30 K. It is believed that BCS theory alone cannot explain this phenomenon and that other effects are in play. These effects are still not yet fully understood; it is possible that they even control superconductivity at low temperatures for some materials.

Overview

At sufficiently low temperatures, electrons near the Fermi surface become unstable against the formation of Cooper pairs. Cooper showed such binding will occur in the presence of an attractive potential, no matter how weak. In conventional superconductors, an attraction is generally attributed to an electron-lattice interaction. The BCS theory, however, requires only that the potential be attractive, regardless of its origin. In the BCS framework, superconductivity is a macroscopic effect which results from the condensation of Cooper pairs. These have some bosonic properties, and bosons, at sufficiently low temperature, can form a large Bose–Einstein condensate. Superconductivity was simultaneously explained by Nikolay Bogolyubov, by means of the Bogoliubov transformations.

In many superconductors, the attractive interaction between electrons (necessary for pairing) is brought about indirectly by the interaction between the electrons and the vibrating crystal lattice (the phonons). Roughly speaking the picture is the following:

An electron moving through a conductor will attract nearby positive charges in the lattice. This deformation of the lattice causes another electron, with opposite spin, to move into the region of higher positive charge density. The two electrons then become correlated. Because there are a lot of such electron pairs in a superconductor, these pairs overlap very strongly and form a highly collective condensate. In this "condensed" state, the breaking of one pair will change the energy of the entire condensate - not just a single electron, or a single pair. Thus, the energy required to break any single pair is related to the energy required to break *all* of the pairs (or more than just two electrons). Because the pairing increases this energy barrier, kicks from oscillating atoms in the conductor (which are small at sufficiently low temperatures) are not enough to affect the condensate as a whole, or any individual "member pair" within the condensate. Thus the electrons stay paired together and resist all kicks, and the electron flow as a whole (the current through the superconductor) will not experience resistance. Thus, the collective behavior of the condensate is a crucial ingredient necessary for superconductivity.

Details

BCS theory starts from the assumption that there is some attraction between electrons, which can overcome the Coulomb repulsion. In most materials (in low temperature superconductors), this attraction is brought about indirectly by the coupling of electrons to the crystal lattice (as explained above). However, the results of BCS theory do *not* depend on the origin of the attractive interaction. For instance, Cooper pairs have been observed in ultracold gases of fermions where a homogeneous magnetic field has been tuned to their Feshbach resonance. The original results of BCS (discussed below) described an s-wave superconducting state, which is the rule among low-temperature superconductors but is not realized in many unconventional superconductors such as the d-wave high-temperature superconductors.

Extensions of BCS theory exist to describe these other cases, although they are insufficient to completely describe the observed features of high-temperature superconductivity.

BCS is able to give an approximation for the quantum-mechanical many-body state of the system of (attractively interacting) electrons inside the metal. This state is now known as the BCS state. In the normal state of a metal, electrons move independently, whereas in the BCS state, they are bound into Cooper pairs by the attractive interaction. The BCS formalism is based on the reduced potential for the electrons' attraction. Within this potential, a variational ansatz for the wave function is proposed. This ansatz was later shown to be exact in the dense limit of pairs. Note that the continuous crossover between the dilute and dense regimes of attracting pairs of fermions is still an open problem, which now attracts a lot of attention within the field of ultracold gases.

Underlying Evidence

The hyperphysics website pages at Georgia State University summarize some key background to BCS theory as follows:

- Evidence of a band gap at the Fermi level (described as "a key piece in the puzzle")

  the existence of a critical temperature and critical magnetic field implied a band gap, and suggested a phase transition, but single electrons are forbidden from condensing to the same energy level by the Pauli exclusion principle. The site comments that "a drastic change in conductivity demanded a drastic change in electron behavior". Conceivably, pairs of electrons might perhaps act like bosons instead, which are bound by different condensate rules and do not have the same limitation.

- Isotope effect on the critical temperature, suggesting lattice interactions

  The Debye frequency of phonons in a lattice is proportional to the inverse of the square root of the mass of lattice ions. It was shown that the superconducting transition temperature of mercury indeed showed the same dependence, by substituting natural mercury ^{202}Hg with a different isotope ^{198}Hg.

- An exponential rise in heat capacity near the critical temperature for some superconductors

  An exponential increase in heat capacity near the critical temperature also suggests an energy bandgap for the superconducting material. As superconducting vanadium is warmed toward its critical temperature, its heat capacity increases massively in a very few degrees; this suggests an energy gap being bridged by thermal energy.

- The lessening of the measured energy gap towards the critical temperature

  This suggests a type of situation where some kind of binding energy exists but it is gradually weakened as the critical temperature is approached. A binding energy suggests two or more particles or other entities that are bound together in the superconducting state. This helped to support the idea of bound particles - specifically electron pairs - and together with the above helped to paint a general picture of paired electrons and their lattice interactions.

Implications

BCS derived several important theoretical predictions that are independent of the details of the interaction, since the quantitative predictions mentioned below hold for any sufficiently weak attraction between the electrons and this last condition is fulfilled for many low temperature superconductors - the so-called weak-coupling case. These have been confirmed in numerous experiments:

- The electrons are bound into Cooper pairs, and these pairs are correlated due to the Pauli exclusion principle for the electrons, from which they are constructed. Therefore, in order to break a pair, one has to change energies of all other pairs. This means there is an energy gap for single-particle excitation, unlike in the normal metal (where the state of an electron can be changed by adding an arbitrarily small amount of energy). This energy gap is highest at low temperatures but vanishes at the transition temperature when superconductivity ceases to exist. The BCS theory gives an expression that shows how the gap grows with the strength of the attractive interaction and the (normal phase) single particle density of states at the Fermi level. Furthermore, it describes how the density of states is changed on entering the superconducting state, where there are no electronic states any more at the Fermi level. The energy gap is most directly observed in tunneling experiments and in reflection of microwaves from superconductors.

- BCS theory predicts the dependence of the value of the energy gap Δ at temperature T on the critical temperature T_c. The ratio between the value of the energy gap at zero temperature and the value of the superconducting transition temperature (expressed in energy units) takes the universal value

$$\Delta(T=0)=1.764k_BT_c,$$

independent of material. Near the critical temperature the relation asymptotes to

$$\Delta(T\to T_c)\approx 3.07k_BT_c\sqrt{1-(T/T_c)}$$

which is of the form suggested the previous year by M. J. Buckingham based on the fact that the superconducting phase transition is second order, that the superconducting phase has a mass gap and on Blevins, Gordy and Fairbank's experimental results the previous year on the absorption of millimeter waves by superconducting tin.

- Due to the energy gap, the specific heat of the superconductor is suppressed strongly (exponentially) at low temperatures, there being no thermal excitations left. However, before reaching the transition temperature, the specific heat of the superconductor becomes even higher than that of the normal conductor (measured immediately above the transition) and the ratio of these two values is found to be universally given by 2.5.

- BCS theory correctly predicts the Meissner effect, i.e. the expulsion of a magnetic field from the superconductor and the variation of the penetration depth (the extent of the screening currents flowing below the metal's surface) with temperature.

- It also describes the variation of the critical magnetic field (above which the superconductor can no longer expel the field but becomes normal conducting) with temperature. BCS theory relates the value of the critical field at zero temperature to the value of the transition temperature and the density of states at the Fermi level.
- In its simplest form, BCS gives the superconducting transition temperature T_c in terms of the electron-phonon coupling potential V and the Debye cutoff energy E_D:

$$k_B T_c = 1.13 E_D e^{-1/N(0)V},$$

where $N(0)$ is the electronic density of states at the Fermi level.

- The BCS theory reproduces the isotope effect, which is the experimental observation that for a given superconducting material, the critical temperature is inversely proportional to the mass of the isotope used in the material. The isotope effect was reported by two groups on 24 March 1950, who discovered it independently working with different mercury isotopes, although a few days before publication they learned of each other's results at the ONR conference in Atlanta. The two groups are Emanuel Maxwell, who published his results in Isotope Effect in the Superconductivity of Mercury and C. A. Reynolds, B. Serin, W. H. Wright, and L. B. Nesbitt who published their results 10 pages later in Superconductivity of Isotopes of Mercury. The choice of isotope ordinarily has little effect on the electrical properties of a material, but does affect the frequency of lattice vibrations. This effect suggests that superconductivity is related to vibrations of the lattice. This is incorporated into BCS theory, where lattice vibrations yield the binding energy of electrons in a Cooper pair.
- Little-Parks experiment - One of the first indications to the importance of the Cooper-pairing principle.

References

- Cross, Michael. "Fermi Liquid Theory: Principles" (PDF). California Institute of Technology. Retrieved 2 February 2015
- C. C. Homes; J. J. Tu; J. Li; G. D. Gu; A. Akrap (2013). "Optical conductivity of nodal metals". SciRep. 3: 3446. Bibcode:2013NatSR...3E3446H. arXiv:1312.4466. doi:10.1038/srep03446
- Lev D. Landau; Evgeny M. Lifschitz (1984). Electrodynamics of Continuous Media. Course of Theoretical Physics. 8. Oxford: Butterworth-Heinemann. ISBN 0-7506-2634-8
- Wysokiński, Carol; et al. (2003). "Spin triplet superconductivity in Sr2RuO4" (PDF). Physica Status Solidi. 236 (2). Bibcode:2003PSSBR.236..325W. arXiv:cond-mat/0211199. doi:10.1002/pssb.200301672. Retrieved 8 April 2012
- Lifshitz, E. M.; Pitaevskii, L.P. (1980). Statistical Physics (Part 2). Landau and Lifshitz. 9. Elsevier. ISBN 0-7506-2636-4
- Abrikosov, A. A. (1957). The magnetic properties of superconducting alloys. Journal of Physics and Chemistry of Solids, 2(3), 199–208
- London, F. (September 1948). "On the Problem of the Molecular Theory of Superconductivity". Physical Review. 74 (5): 562–573. Bibcode:1948PhRv...74..562L. doi:10.1103/PhysRev.74.562. Retrieved March 3, 2012
- Katsuda, A.; Sunada, T (1987). "Homology and closed geodesics in a compact Riemann surface". Amer. J. Math. 110: 145–156. doi:10.2307/2374542

- Maxwell, Emanuel. "Isotope Effect in the Superconductivity of Mercury". Physical Review. 78 (4): 477–477. Bibcode:1950PhRv...78..477M. doi:10.1103/PhysRev.78.477. Retrieved 27 July 2014
- Ong, edited by N. Phuan; Bhatt, Ravin N. (2001). More is different : fifty years of condensed matter physics. Princeton (N.J.): Princeton university press. p. 65. ISBN 0691088667. Retrieved 2 February 2015
- Bardeen, J.; Cooper, L. N.; Schrieffer, J. R. (December 1957). "Theory of Superconductivity". Physical Review. 108 (5): 1175–1204. Bibcode:1957PhRv..108.1175B. doi:10.1103/PhysRev.108.1175. Retrieved May 3, 2012
- Buckingham, M. J. (February 1956). "Very High Frequency Absorption in Superconductors". Physical Review. 101: 1431–1432. Bibcode:1956PhRv..101.1431B. doi:10.1103/PhysRev.101.1431

Understanding Crystallography

Crystallography studies the arrangement of atoms in crystalline solids. X-rays, electron and neutron beams are used to study crystalline structure. The aspects elucidated in this chapter are of vital importance, and provide a better understanding of condensed matter theory.

Crystallography

A crystalline solid: atomic resolution image of strontium titanate.
Brighter atoms are strontium and darker ones are titanium.

Crystallography is the experimental science of determining the arrangement of atoms in the crystalline solids. The word "crystallography" derives from the Greek words *crystallon* "cold drop, frozen drop", with its meaning extending to all solids with some degree of transparency, and *graphein* "to write". In July 2012, the United Nations recognised the importance of the science of crystallography by proclaiming that 2014 would be the International Year of Crystallography. X-ray crystallography is used to determine the structure of large biomolecules such as proteins. Before the development of X-ray diffraction crystallography, the study of crystals was based on physical measurements of their geometry. This involved measuring the angles of crystal faces relative to each other and to theoretical reference axes (crystallographic axes), and establishing the symmetry of the crystal in question. This physical measurement is carried out using a goniometer. The position in 3D space of each crystal face is plotted on a stereographic net such as a Wulff net or Lambert net. The pole to each face is plotted on the net. Each point is labelled with its Miller index. The final plot allows the symmetry of the crystal to be established.

Crystallographic methods now depend on analysis of the diffraction patterns of a sample targeted by a beam of some type. X-rays are most commonly used; other beams used include electrons or neutrons. This is facilitated by the wave properties of the particles. Crystallographers often explicitly state the type of beam used, as in the terms *X-ray crystallography, neutron diffraction* and *electron diffraction.* These three types of radiation interact with the specimen in different ways.

- X-rays interact with the spatial distribution of electrons in the sample.
- Electrons are charged particles and therefore interact with the total charge distribution of both the atomic nuclei and the electrons of the sample.
- Neutrons are scattered by the atomic nuclei through the strong nuclear forces, but in addition, the magnetic moment of neutrons is non-zero. They are therefore also scattered by magnetic fields. When neutrons are scattered from hydrogen-containing materials, they produce diffraction patterns with high noise levels. However, the material can sometimes be treated to substitute deuterium for hydrogen.

Because of these different forms of interaction, the three types of radiation are suitable for different crystallographic studies.

Theory

An image of a small object is made using a lens to focus the beam, similar to a lens in a microscope. However, the wavelength of visible light (about 4000 to 7000 ångström) is three orders of magnitude longer than the length of typical atomic bonds and atoms themselves (about 1 to 2 Å). Therefore, obtaining information about the spatial arrangement of atoms requires the use of radiation with shorter wavelengths, such as X-ray or neutron beams. Employing shorter wavelengths implied abandoning microscopy and true imaging, however, because there exists no material from which a lens capable of focusing this type of radiation can be created. (Nevertheless, scientists have had some success focusing X-rays with microscopic Fresnel zone plates made from gold, and by critical-angle reflection inside long tapered capillaries.) Diffracted X-ray or neutron beams cannot be focused to produce images, so the sample structure must be reconstructed from the diffraction pattern. Sharp features in the diffraction pattern arise from periodic, repeating structure in the sample, which are often very strong due to coherent reflection of many photons from many regularly spaced instances of similar structure, while non-periodic components of the structure result in diffuse (and usually weak) diffraction features - areas with a higher density and repetition of atom order tend to reflect more light toward one point in space when compared to those areas with fewer atoms and less repetition.

Because of their highly ordered and repetitive structure, crystals give diffraction patterns of sharp Bragg reflection spots, and are ideal for analyzing the structure of solids.

Notation

- Coordinates in *square brackets* such as denote a direction vector (in real space).
- Coordinates in *angle brackets* or *chevrons* such as <100> denote a *family* of directions which are related by symmetry operations. In the cubic crystal system for example, <100> would mean , , or the negative of any of those directions.

- Miller indices in *parentheses* such as (100) denote a plane of the crystal structure, and regular repetitions of that plane with a particular spacing. In the cubic system, the normal to the (hkl) plane is the direction [hkl], but in lower-symmetry cases, the normal to (hkl) is not parallel to [hkl].
- Indices in *curly brackets* or *braces* such as {100} denote a family of planes and their normals which are equivalent in cubic materials due to symmetry operations, much the way angle brackets denote a family of directions. In non-cubic materials, <hkl> is not necessarily perpendicular to {hkl}.

Techniques

Some materials that have been analyzed crystallographically, such as proteins, do not occur naturally as crystals. Typically, such molecules are placed in solution and allowed to slowly crystallize through vapor diffusion. A drop of solution containing the molecule, buffer, and precipitants is sealed in a container with a reservoir containing a hygroscopic solution. Water in the drop diffuses to the reservoir, slowly increasing the concentration and allowing a crystal to form. If the concentration were to rise more quickly, the molecule would simply precipitate out of solution, resulting in disorderly granules rather than an orderly and hence usable crystal.

Once a crystal is obtained, data can be collected using a beam of radiation. Although many universities that engage in crystallographic research have their own X-ray producing equipment, synchrotrons are often used as X-ray sources, because of the purer and more complete patterns such sources can generate. Synchrotron sources also have a much higher intensity of X-ray beams, so data collection takes a fraction of the time normally necessary at weaker sources. Complementary neutron crystallography techniques are used to identify the positions of hydrogen atoms, since X-rays only interact very weakly with light elements such as hydrogen.

Producing an image from a diffraction pattern requires sophisticated mathematics and often an iterative process of modelling and refinement. In this process, the mathematically predicted diffraction patterns of an hypothesized or "model" structure are compared to the actual pattern generated by the crystalline sample. Ideally, researchers make several initial guesses, which through refinement all converge on the same answer. Models are refined until their predicted patterns match to as great a degree as can be achieved without radical revision of the model. This is a painstaking process, made much easier today by computers.

The mathematical methods for the analysis of diffraction data only apply to *patterns,* which in turn result only when waves diffract from orderly arrays. Hence crystallography applies for the most part only to crystals, or to molecules which can be coaxed to crystallize for the sake of measurement. In spite of this, a certain amount of molecular information can be deduced from patterns that are generated by fibers and powders, which while not as perfect as a solid crystal, may exhibit a degree of order. This level of order can be sufficient to deduce the structure of simple molecules, or to determine the coarse features of more complicated molecules. For example, the double-helical structure of DNA was deduced from an X-ray diffraction pattern that had been generated by a fibrous sample.

In Materials Science

Crystallography is used by materials scientists to characterize different materials. In single crystals, the effects of the crystalline arrangement of atoms is often easy to see macroscopically, because the natural shapes of crystals reflect the atomic structure. In addition, physical properties are often controlled by crystalline defects. The understanding of crystal structures is an important prerequisite for understanding crystallographic defects. Mostly, materials do not occur as a single crystal, but in poly-crystalline form (i.e., as an aggregate of small crystals with different orientations). Because of this, the powder diffraction method, which takes diffraction patterns of polycrystalline samples with a large number of crystals, plays an important role in structural determination.

Other physical properties are also linked to crystallography. For example, the minerals in clay form small, flat, platelike structures. Clay can be easily deformed because the platelike particles can slip along each other in the plane of the plates, yet remain strongly connected in the direction perpendicular to the plates. Such mechanisms can be studied by crystallographic texture measurements.

In another example, iron transforms from a body-centered cubic (bcc) structure to a face-centered cubic (fcc) structure called austenite when it is heated. The fcc structure is a close-packed structure unlike the bcc structure; thus the volume of the iron decreases when this transformation occurs.

Crystallography is useful in phase identification. When manufacturing or using a material, it is generally desirable to know what compounds and what phases are present in the material, as their composition, structure and proportions will influence the material's properties. Each phase has a characteristic arrangement of atoms. X-ray or neutron diffraction can be used to identify which patterns are present in the material, and thus which compounds are present. Crystallography covers the enumeration of the symmetry patterns which can be formed by atoms in a crystal and for this reason is related to group theory and geometry.

Biology

X-ray crystallography is the primary method for determining the molecular conformations of biological macromolecules, particularly protein and nucleic acids such as DNA and RNA. In fact, the double-helical structure of DNA was deduced from crystallographic data. The first crystal structure of a macromolecule was solved in 1958, a three-dimensional model of the myoglobin molecule obtained by X-ray analysis. The Protein Data Bank (PDB) is a freely accessible repository for the structures of proteins and other biological macromolecules. Computer programs such as RasMol or Pymol can be used to visualize biological molecular structures. Neutron crystallography is often used to help refine structures obtained by X-ray methods or to solve a specific bond; the methods are often viewed as complementary, as X-rays are sensitive to electron positions and scatter most strongly off heavy atoms, while neutrons are sensitive to nucleus positions and scatter strongly even off many light isotopes, including hydrogen and deuterium. Electron crystallography has been used to determine some protein structures, most notably membrane proteins and viral capsids.

Crystal Structure

In crystallography, crystal structure is a description of the ordered arrangement of atoms, ions or molecules in a crystalline material. Ordered structures occur from the intrinsic nature of the constituent particles to form symmetric patterns that repeat along the principal directions of three-dimensional space in matter.

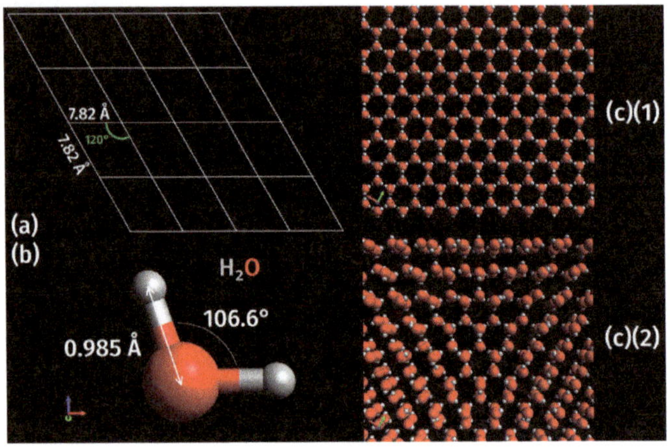

The (3-D) crystal structure of H_2O ice Ih (c) consists of bases of H_2O ice molecules (b) located on lattice points within the (2-D) hexagonal space lattice (a). The values for the H–O–H angle and O–H distance have come from *Physics of Ice* with uncertainties of ±1.5° and ±0.005 Å, respectively. The white box in (c) is the unit cell defined by Bernal and Fowler

The smallest group of particles in the material that constitutes the repeating pattern is the unit cell of the structure. The unit cell completely defines the symmetry and structure of the entire crystal lattice, which is built up by repetitive translation of the unit cell along its principal axes. The repeating patterns are said to be located at the points of the Bravais lattice.

The lengths of the principal axes, or edges, of the unit cell and the angles between them are the lattice constants, also called *lattice parameters*. The symmetry properties of the crystal are described by the concept of space groups. All possible symmetric arrangements of particles in three-dimensional space may be described by the 230 space groups.

The crystal structure and symmetry play a critical role in determining many physical properties, such as cleavage, electronic band structure, and optical transparency.

Unit Cell

The crystal structure of a material (the arrangement of atoms within a given type of crystal) can be described in terms of its unit cell. The unit cell is a box containing one or more atoms arranged in three dimensions. The unit cells stacked in three-dimensional space describe the bulk arrangement of atoms of the crystal. The unit cell is represented in terms of its lattice parameters, which are the lengths of the cell edges (a, b and c) and the angles between them (alpha, beta and gamma), while the positions of the atoms inside the unit cell are described by the set of atomic positions (x_i, y_i, z_i) measured from a reference lattice point. Commonly, atomic positions are represented in terms of fractional coordinates, relative to the unit cell lengths.

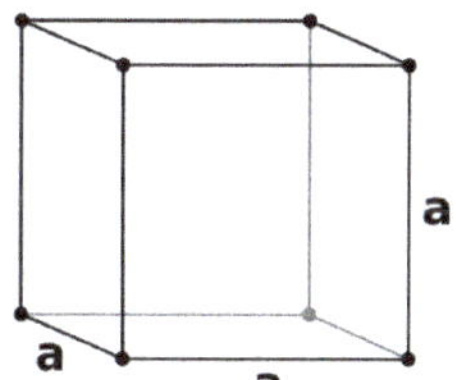

Simple cubic (P)

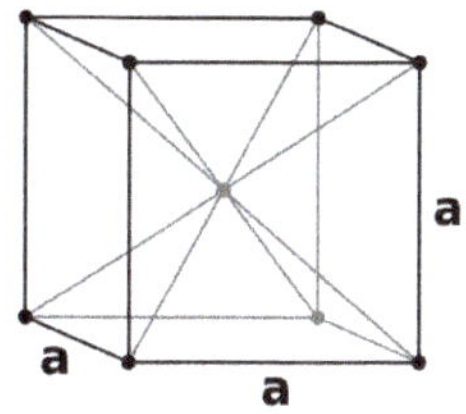

Body-centered cubic (I)

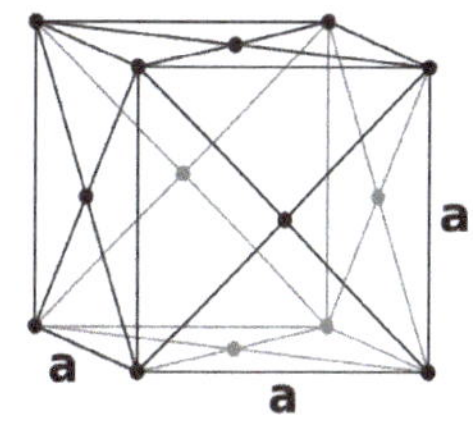

Face-centered cubic (F)

The atom positions within the unit cell can be calculated through application of symmetry operations to the asymmetric unit. The asymmetric unit refers to the smallest possible occupation of space within the unit cell. This does not, however imply that the entirety of the asymmetric unit must lie within the boundaries of the unit cell. Symmetric transformations of atom positions are calculated from the space group of the crystal structure, and this is usually a black box operation performed by computer programs. However, manual calculation of the atomic positions within the unit cell can be performed from the asymmetric unit, through the application of the symmetry operators described within the *International Tables for Crystallography: Volume A.*

Miller Indices

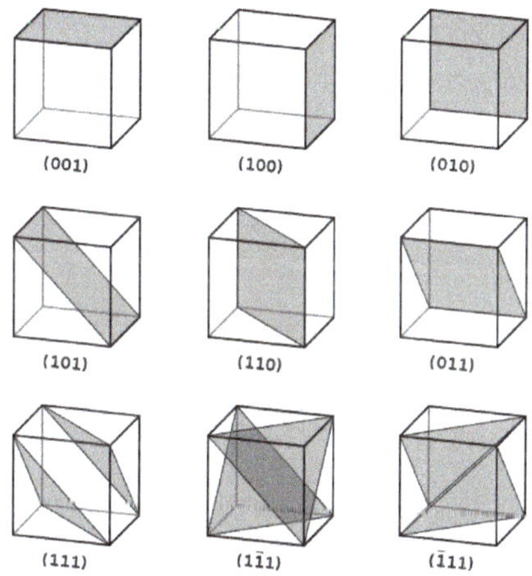

Planes with different Miller indices in cubic crystals

Vectors and planes in a crystal lattice are described by the three-value Miller index notation. It uses the indices ℓ, m, and n as directional parameters, which are separated by 90°, and are thus orthogonal.

By definition, the syntax (ℓmn) denotes a plane that intercepts the three points a_1/ℓ, a_2/m, and a_3/n, or some multiple thereof. That is, the Miller indices are proportional to the inverses of the intercepts of the plane with the unit cell (in the basis of the lattice vectors). If one or more of the indices is zero, it means that the planes do not intersect that axis (i.e., the intercept is "at infinity"). A plane containing a coordinate axis is translated so that it no longer contains that axis before its Miller indices are determined. The Miller indices for a plane are integers with no common factors. Negative indices are indicated with horizontal bars, as in (123). In an orthogonal coordinate system for a cubic cell, the Miller indices of a plane are the Cartesian components of a vector normal to the plane.

Considering only (ℓmn) planes intersecting one or more lattice points (the *lattice planes*), the distance d between adjacent lattice planes is related to the (shortest) reciprocal lattice vector orthogonal to the planes by the formula

$$d = \frac{2\pi}{|\mathbf{g}_{\ell mn}|}$$

Planes and Directions

The crystallographic directions are geometric lines linking nodes (atoms, ions or molecules) of a crystal. Likewise, the crystallographic planes are geometric *planes* linking nodes. Some directions and planes have a higher density of nodes. These high density planes have an influence on the behavior of the crystal as follows:

- Optical properties: Refractive index is directly related to density (or periodic density fluctuations).
- Adsorption and reactivity: Physical adsorption and chemical reactions occur at or near surface atoms or molecules. These phenomena are thus sensitive to the density of nodes.
- Surface tension: The condensation of a material means that the atoms, ions or molecules are more stable if they are surrounded by other similar species. The surface tension of an interface thus varies according to the density on the surface.

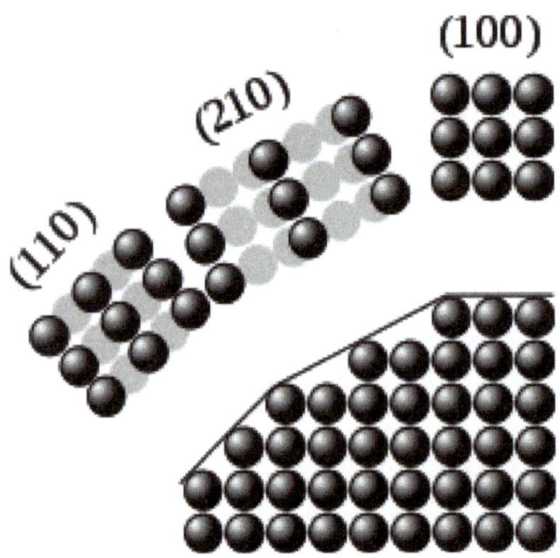

Dense crystallographic planes

- Microstructural defects: Pores and crystallites tend to have straight grain boundaries following higher density planes.
- Cleavage: This typically occurs preferentially parallel to higher density planes.
- Plastic deformation: Dislocation glide occurs preferentially parallel to higher density planes. The perturbation carried by the dislocation (Burgers vector) is along a dense direction. The shift of one node in a more dense direction requires a lesser distortion of the crystal lattice.

Some directions and planes are defined by symmetry of the crystal system. In monoclinic, rhombohedral, tetragonal, and trigonal/hexagonal systems there is one unique axis (sometimes called the principal axis) which has higher rotational symmetry than the other two axes. The basal plane is the plane perpendicular to the principal axis in these crystal systems. For triclinic, orthorhombic, and cubic crystal systems the axis designation is arbitrary and there is no principal axis.

Cubic Structures

For the special case of simple cubic crystals, the lattice vectors are orthogonal and of equal length (usually denoted *a*); similarly for the reciprocal lattice. So, in this common case, the Miller indices (ℓmn) and [ℓmn] both simply denote normals/directions in Cartesian coordinates. For cubic crystals with lattice constant *a*, the spacing *d* between adjacent (ℓmn) lattice planes is (from above):

$$d_{\ell mn} = \frac{a}{\sqrt{\ell^2 + m^2 + n^2}}$$

Because of the symmetry of cubic crystals, it is possible to change the place and sign of the integers and have equivalent directions and planes:

- Coordinates in *angle brackets* such as ⟨100⟩ denote a *family* of directions that are equivalent due to symmetry operations, such as [100], [010], [001] or the negative of any of those directions.
- Coordinates in *curly brackets* or *braces* such as {100} denote a family of plane normals that are equivalent due to symmetry operations, much the way angle brackets denote a family of directions.

For face-centered cubic (fcc) and body-centered cubic (bcc) lattices, the primitive lattice vectors are not orthogonal. However, in these cases the Miller indices are conventionally defined relative to the lattice vectors of the cubic supercell and hence are again simply the Cartesian directions.

Interplanar Spacing

The spacing *d* between adjacent (*hkl*) lattice planes is given by:

- Cubic:

$$\frac{1}{d^2} = \frac{h^2 + k^2 + l^2}{a^2}$$

- Tetragonal:

$$\frac{1}{d^2} = \frac{h^2 + k^2}{a^2} + \frac{l^2}{c^2}$$

- Hexagonal:

$$\frac{1}{d^2} = \frac{4}{3}\left(\frac{h^2 + hk + k^2}{a^2}\right) + \frac{l^2}{c^2}$$

- Rhombohedral:

$$\frac{1}{d^2} = \frac{(h^2 + k^2 + l^2)\sin^2\alpha + 2(hk + kl + hl)(\cos^2\alpha - \cos\alpha)}{a^2(1 - 3\cos^2\alpha + 2\cos^3\alpha)}$$

- Orthorhombic:

$$\frac{1}{d^2} = \frac{h^2}{a^2} + \frac{k^2}{b^2} + \frac{l^2}{c^2}$$

- Monoclinic:

$$\frac{1}{d^2} = \left(\frac{h^2}{a^2} + \frac{k^2\sin^2\beta}{b^2} + \frac{l^2}{c^2} - \frac{2hl\cos\beta}{ac}\right)\csc^2\beta$$

- Triclinic:

$$\frac{1}{d^2} = \frac{\frac{h^2}{a^2}\sin^2\alpha + \frac{k^2}{b^2}\sin^2\beta + \frac{l^2}{c^2}\sin^2\gamma + \frac{2kl}{bc}\cos\alpha + \frac{2hl}{ac}\cos\beta + \frac{2hk}{ab}\cos\gamma}{1 - \cos^2\alpha - \cos^2\beta - \cos^2\gamma + 2\cos\alpha\cos\beta\cos\gamma}$$

Classification by Symmetry

The defining property of a crystal is its inherent symmetry, by which we mean that under certain 'operations' the crystal remains unchanged. All crystals have translational symmetry in three directions, but some have other symmetry elements as well. For example, rotating the crystal 180° about a certain axis may result in an atomic configuration that is identical to the original configuration. The crystal is then said to have a twofold rotational symmetry about this axis. In addition to rotational symmetries like this, a crystal may have symmetries in the form of mirror planes and translational symmetries, and also the so-called "compound symmetries," which are a combination of translation and rotation/mirror symmetries. A full classification of a crystal is achieved when all of these inherent symmetries of the crystal are identified.

Lattice Systems

These lattice systems are a grouping of crystal structures according to the axial system used to describe their lattice. Each lattice system consists of a set of three axes in a particular geometric arrangement. There are seven lattice systems. They are similar to but not quite the same as the seven crystal systems.

Crystal Family	Lattice System	Schönflies	14 Bravais Lattices			
			Primitive	Base-centered	Body-centered	Face-centered
triclinic		C_i	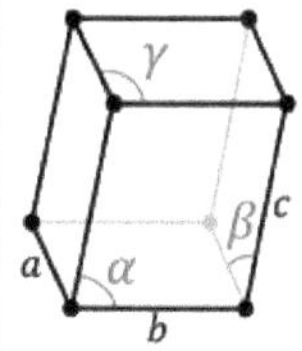			
monoclinic		C_{2h}	$\beta \neq 90°$ $a \neq c$ 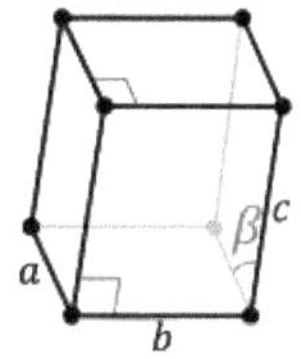	$\beta \neq 90°$ $a \neq c$ 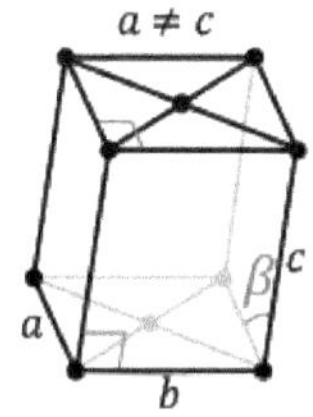		
orthorhombic		D_{2h}	$a \neq b \neq c$ 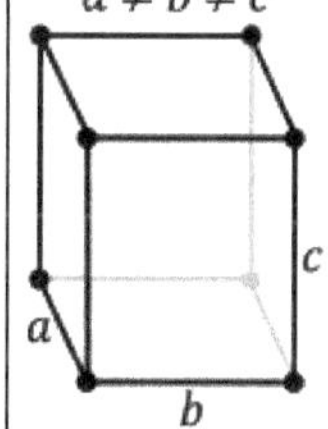	$a \neq b \neq c$ 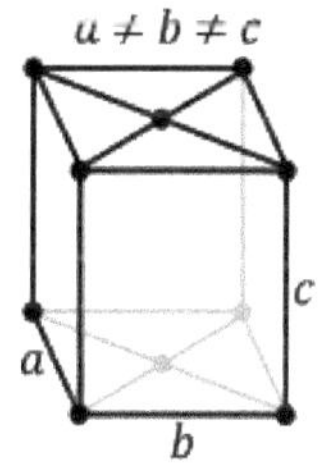	$a \neq b \neq c$ 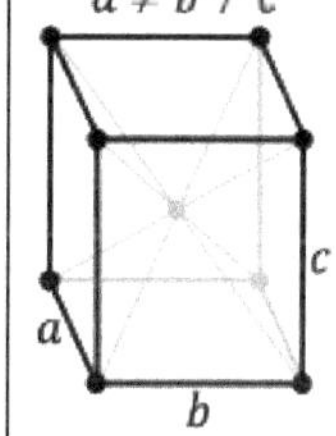	$a \neq b \neq c$ 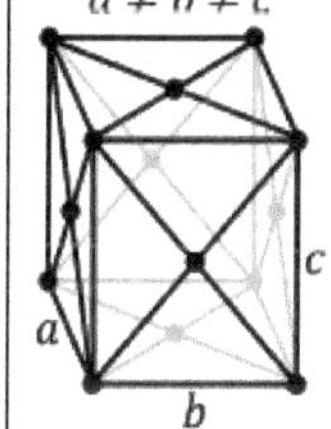
tetragonal		D_{4h}	$a \neq c$ 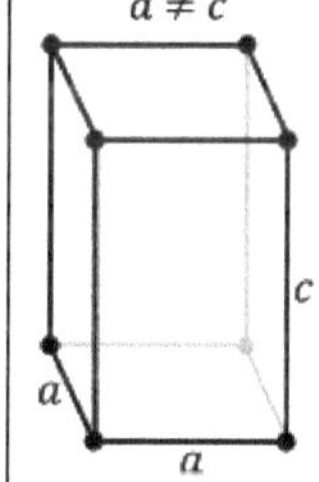		$a \neq c$ 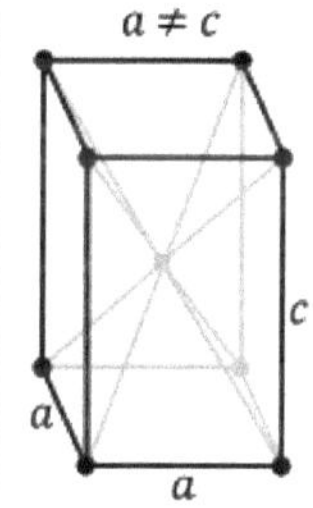	
hexagonal	rhombohedral	D_{3d}	$\alpha \neq 90°$ 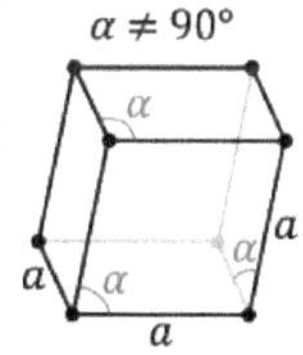			
	hexagonal	D_{6h}	$\gamma = 120°$ 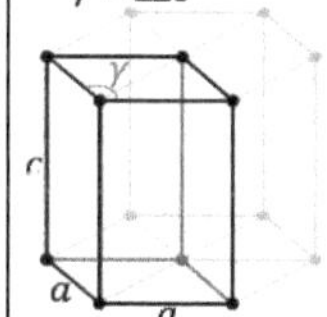			
cubic		O_h	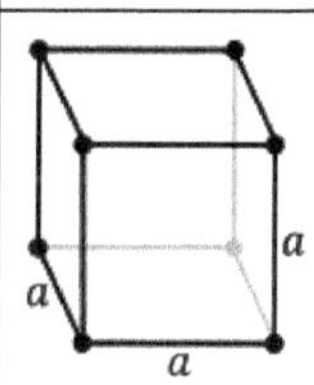		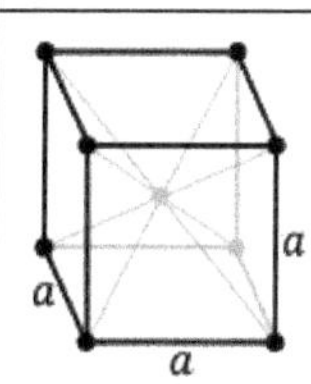	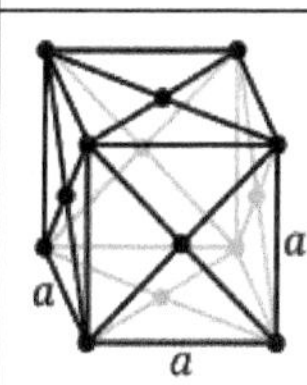

The simplest and most symmetric, the cubic (or isometric) system, has the symmetry of a cube, that is, it exhibits four threefold rotational axes oriented at 109.5° (the tetrahedral angle) with respect to each other. These threefold axes lie along the body diagonals of the cube. The other six lattice systems, are hexagonal, tetragonal, rhombohedral (often confused with the trigonal crystal system), orthorhombic, monoclinic and triclinic.

Bravais Lattices

Bravais lattices, also referred to as *space lattices*, describe the geometric arrangement of the lattice points, and therefore the translational symmetry of the crystal. The three dimensions of space afford 14 distinct Bravais lattices describing the translational symmetry. All crystalline materials recognized today, not including quasicrystals, fit in one of these arrangements. The fourteen three-dimensional lattices, classified by lattice system, are shown.

The crystal structure consists of the same group of atoms, the *basis*, positioned around each and every lattice point. This group of atoms therefore repeats indefinitely in three dimensions according to the arrangement of one of the Bravais lattices. The characteristic rotation and mirror symmetries of the unit cell is described by its crystallographic point group.

Crystal Systems

A crystal system is a set of point groups in which the point groups themselves and their corresponding space groups are assigned to a lattice system. Of the 32 point groups that exist in three dimensions, most are assigned to only one lattice system, in which case the crystal system and lattice system both have the same name. However, five point groups are assigned to two lattice systems, rhombohedral and hexagonal, because both lattice systems exhibit threefold rotational symmetry. These point groups are assigned to the trigonal crystal system.

Crystal Family	**Crystal System**	**Point Group / Crystal Class**	**Schönflies**	**Point Symmetry**	**Order**	**Abstract Group**
triclinic		triclinic-pedial	C_1	enantiomorphic polar	1	trivial
triclinic-pinacoidal		C_i	centrosymmetric	2	cyclic	
monoclinic		monoclinic-sphenoidal	C_2	enantiomorphic polar	2	cyclic
monoclinic-domatic		C_s	polar	2	cyclic	
monoclinic-prismatic		C_{2h}	centrosymmetric	4	Klein four	
orthorhombic		orthorhombic-sphenoidal	D_2	enantiomorphic	4	Klein four
orthorhombic-pyramidal		C_{2v}	polar	4	Klein four	
orthorhombic-bipyramidal		D_{2h}	centrosymmetric	8		

tetragonal		tetragonal-pyramidal	C_4	enantiomorphic polar	4	cyclic
tetragonal-disphenoidal		S_4	non-centrosymmetric	4	cyclic	
tetragonal-dipyramidal		C_{4h}	centrosymmetric	8		
tetragonal-trapezoidal		D_4	enantiomorphic	8	dihedral	
ditetragonal-pyramidal		C_{4v}	polar	8	dihedral	
tetragonal-scalenoidal		D_{2d}	non-centrosymmetric	8	dihedral	
ditetragonal-dipyramidal		D_{4h}	centrosymmetric	16		
hexagonal	trigonal	trigonal-pyramidal	C_3	enantiomorphic polar	3	cyclic
		rhombohedral	S_6 (C_{3i})	centrosymmetric	6	cyclic
		trigonal-trapezoidal	D_3	enantiomorphic	6	dihedral
		ditrigonal-pyramidal	C_{3v}	polar	6	dihedral
		ditrigonal-scalahedral	D_{3d}	centrosymmetric	12	dihedral
	hexagonal	hexagonal-pyramidal	C_6	enantiomorphic polar	6	cyclic
		trigonal-dipyramidal	C_{3h}	non-centrosymmetric	6	cyclic
		hexagonal-dipyramidal	C_{6h}	centrosymmetric	12	
		hexagonal-trapezoidal	D_6	enantiomorphic	12	dihedral
		dihexagonal-pyramidal	C_{6v}	polar	12	dihedral
		ditrigonal-dipyramidal	D_{3h}	non-centrosymmetric	12	dihedral
		dihexagonal-dipyramidal	D_{6h}	centrosymmetric	24	
cubic		tetrahedral	T	enantiomorphic	12	alternating
hextetrahedral		T_d	non-centrosymmetric	24	symmetric	
diploidal		T_h	centrosymmetric	24		
gyroidal		O	enantiomorphic	24	symmetric	
hexoctahedral		O_h	centrosymmetric	48		

In total there are seven crystal systems: triclinic, monoclinic, orthorhombic, tetragonal, trigonal, hexagonal, and cubic.

Point Groups

The crystallographic point group or *crystal class* is the mathematical group comprising the symmetry operations that leave at least one point unmoved and that leave the appearance of the crystal structure unchanged. These symmetry operations include

- *Reflection*, which reflects the structure across a *reflection plane*
- *Rotation*, which rotates the structure a specified portion of a circle about a *rotation axis*
- *Inversion*, which changes the sign of the coordinate of each point with respect to a *center of symmetry* or *inversion point*
- *Improper rotation*, which consists of a rotation about an axis followed by an inversion.

Rotation axes (proper and improper), reflection planes, and centers of symmetry are collectively called *symmetry elements*. There are 32 possible crystal classes. Each one can be classified into one of the seven crystal systems.

Space Groups

In addition to the operations of the point group, the space group of the crystal structure contains translational symmetry operations. These include:

- Pure *translations*, which move a point along a vector
- *Screw axes*, which rotate a point around an axis while translating parallel to the axis.
- *Glide planes*, which reflect a point through a plane while translating it parallel to the plane.

There are 230 distinct space groups.

Atomic Coordination

By considering the arrangement of atoms relative to each other, their coordination numbers (or number of nearest neighbors), interatomic distances, types of bonding, etc., it is possible to form a general view of the structures and alternative ways of visualizing them.

Close Packing

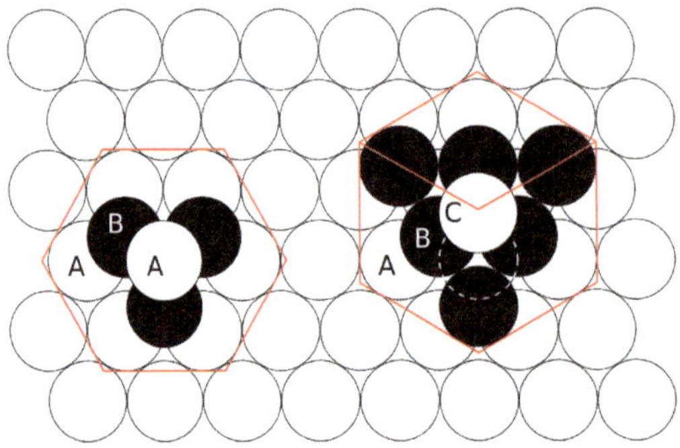

The hcp lattice (left) and the fcc lattice (right)

The principles involved can be understood by considering the most efficient way of packing together equal-sized spheres and stacking close-packed atomic planes in three dimensions. For example, if plane A lies beneath plane B, there are two possible ways of placing an additional atom on top of layer B. If an additional layer was placed directly over plane A, this would give rise to the following series:

...ABABABAB...

This arrangement of atoms in a crystal structure is known as hexagonal close packing (hcp).

If, however, all three planes are staggered relative to each other and it is not until the fourth layer is positioned directly over plane A that the sequence is repeated, then the following sequence arises:

...ABCABCABC...

This type of structural arrangement is known as cubic close packing (ccp).

The unit cell of a ccp arrangement of atoms is the face-centered cubic (fcc) unit cell. This is not immediately obvious as the closely packed layers are parallel to the {111} planes of the fcc unit cell. There are four different orientations of the close-packed layers.

The packing efficiency can be worked out by calculating the total volume of the spheres and dividing by the volume of the cell as follows:

$$\frac{4\times\frac{4}{3}\pi r^3}{16\sqrt{2}r^3}=\frac{\pi}{3\sqrt{2}}=0.7405...$$

The 74% packing efficiency is the maximum density possible in unit cells constructed of spheres of only one size. Most crystalline forms of metallic elements are hcp, fcc, or bcc (body-centered cubic). The coordination number of atoms in hcp and fcc structures is 12 and its atomic packing factor (APF) is the number mentioned above, 0.74. This can be compared to the APF of a bcc structure, which is 0.68.

Grain Boundaries

Grain boundaries are interfaces where crystals of different orientations meet. A grain boundary is a single-phase interface, with crystals on each side of the boundary being identical except in orientation. The term "crystallite boundary" is sometimes, though rarely, used. Grain boundary areas contain those atoms that have been perturbed from their original lattice sites, dislocations, and impurities that have migrated to the lower energy grain boundary.

Treating a grain boundary geometrically as an interface of a single crystal cut into two parts, one of which is rotated, we see that there are five variables required to define a grain boundary. The first two numbers come from the unit vector that specifies a rotation axis. The third number designates the angle of rotation of the grain. The final two numbers specify the plane of the grain boundary (or a unit vector that is normal to this plane).

Grain boundaries disrupt the motion of dislocations through a material, so reducing crystallite size is a common way to improve strength, as described by the Hall–Petch relationship. Since grain boundaries are defects in the crystal structure they tend to decrease the electrical and thermal conductivity of the material. The high interfacial energy and relatively weak bonding in most grain boundaries often makes them preferred sites for the onset of corrosion and for the precipitation of new phases from the solid. They are also important to many of the mechanisms of creep.

Grain boundaries are in general only a few nanometers wide. In common materials, crystallites are large enough that grain boundaries account for a small fraction of the material. However, very small grain sizes are achievable. In nanocrystalline solids, grain boundaries become a significant volume fraction of the material, with profound effects on such properties as diffusion and plasticity. In the limit of small crystallites, as the volume fraction of grain boundaries approaches 100%, the material ceases to have any crystalline character, and thus becomes an amorphous solid.

Defects and Impurities

Real crystals feature defects or irregularities in the ideal arrangements described above and it is these defects that critically determine many of the electrical and mechanical properties of real materials. When one atom substitutes for one of the principal atomic components within the crystal structure, alteration in the electrical and thermal properties of the material may ensue. Impurities may also manifest as spin impurities in certain materials. Research on magnetic impurities demonstrates that substantial alteration of certain properties such as specific heat may be affected by small concentrations of an impurity, as for example impurities in semiconducting ferromagnetic alloys may lead to different properties as first predicted in the late 1960s. Dislocations in the crystal lattice allow shear at lower stress than that needed for a perfect crystal structure.

Prediction of Structure

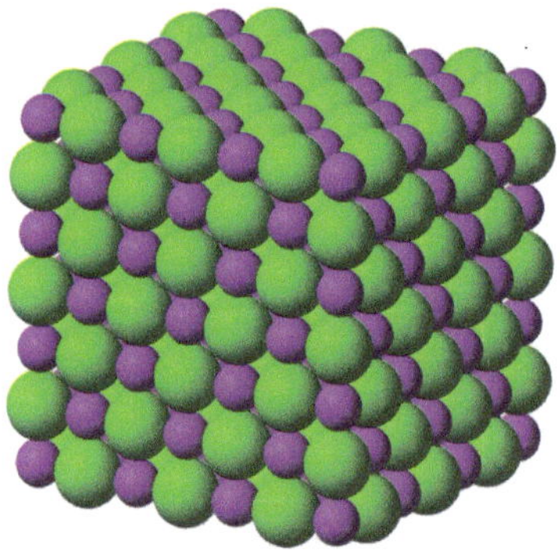

Crystal structure of sodium chloride (table salt)

The difficulty of predicting stable crystal structures based on the knowledge of only the chemical composition has long been a stumbling block on the way to fully computational materials design. Now, with more powerful algorithms and high-performance computing, structures of medium complexity can be predicted using such approaches as evolutionary algorithms, random sampling, or metadynamics.

The crystal structures of simple ionic solids (e.g., NaCl or table salt) have long been rationalized in terms of Pauling's rules, first set out in 1929 by Linus Pauling, referred to by many since as the "father of the chemical bond". Pauling also considered the nature of the interatomic forces in metals, and concluded that about half of the five d-orbitals in the transition metals are involved in bonding, with the remaining nonbonding d-orbitals being responsible for the magnetic properties.

He, therefore, was able to correlate the number of d-orbitals in bond formation with the bond length as well as many of the physical properties of the substance. He subsequently introduced the metallic orbital, an extra orbital necessary to permit uninhibited resonance of valence bonds among various electronic structures.

In the resonating valence bond theory, the factors that determine the choice of one from among alternative crystal structures of a metal or intermetallic compound revolve around the energy of resonance of bonds among interatomic positions. It is clear that some modes of resonance would make larger contributions (be more mechanically stable than others), and that in particular a simple ratio of number of bonds to number of positions would be exceptional. The resulting principle is that a special stability is associated with the simplest ratios or "bond numbers": $1/2$, $1/3$, $2/3$, $1/4$, $3/4$, etc. The choice of structure and the value of the axial ratio (which determines the relative bond lengths) are thus a result of the effort of an atom to use its valency in the formation of stable bonds with simple fractional bond numbers.

After postulating a direct correlation between electron concentration and crystal structure in beta-phase alloys, Hume-Rothery analyzed the trends in melting points, compressibilities and bond lengths as a function of group number in the periodic table in order to establish a system of valencies of the transition elements in the metallic state. This treatment thus emphasized the increasing bond strength as a function of group number. The operation of directional forces were emphasized on the relation between bond hybrids and the metallic structures. The resulting correlation between electronic and crystalline structures is summarized by a single parameter, the weight of the d-electrons per hybridized metallic orbital. The "d-weight" calculates out to 0.5, 0.7 and 0.9 for the fcc, hcp and bcc structures respectively. The relationship between d-electrons and crystal structure thus becomes apparent.

In crystal structure predictions/simulations, the periodicity is usually applied, since the system is imagined as unlimited big in all directions. Starting from a triclinic structure with no further symmetry property assumed, the system may be driven to show some additional symmetry properties by applying Newton's Second Law on particles in the unit cell and a recently developed dynamical equation for the system period vectors (lattice parameters including angles), even if the system is subject to external stress.

Polymorphism

Quartz is one of the several thermodynamically stable crystalline forms of silica, SiO_2. The most important forms of silica include: α-quartz, β-quartz, tridymite, cristobalite, coesite, and stishovite.

Polymorphism is the occurrence of multiple crystalline forms of a material. It is found in many crystalline materials including polymers, minerals, and metals. According to Gibbs' rules of phase equilibria, these unique crystalline phases are dependent on intensive variables such as pressure and temperature. Polymorphism is related to allotropy, which refers to elemental solids. The complete morphology of a material is described by polymorphism and other variables such as crystal habit, amorphous fraction or crystallographic defects. Polymorphs have different stabilities and may spontaneously convert from a metastable form (or thermodynamically unstable form) to the stable form at a particular temperature. They also exhibit different melting points, solubilities, and X-ray diffraction patterns.

One good example of this is the quartz form of silicon dioxide, or SiO_2. In the vast majority of silicates, the Si atom shows tetrahedral coordination by 4 oxygens. All but one of the crystalline forms involve tetrahedral $\{SiO_4\}$ units linked together by shared vertices in different arrangements. In different minerals the tetrahedra show different degrees of networking and polymerization. For example, they occur singly, joined together in pairs, in larger finite clusters including rings, in chains, double chains, sheets, and three-dimensional frameworks. The minerals are classified into groups based on these structures. In each of its 7 thermodynamically stable crystalline forms or polymorphs of crystalline quartz, only 2 out of 4 of each the edges of the $\{SiO_4\}$ tetrahedra are shared with others, yielding the net chemical formula for silica: SiO_2.

Another example is elemental tin (Sn), which is malleable near ambient temperatures but is brittle when cooled. This change in mechanical properties due to existence of its two major allotropes, α- and β-tin. The two allotropes that are encountered at normal pressure and temperature, α-tin and β-tin, are more commonly known as *gray tin* and *white tin* respectively. Two more allotropes, γ and σ, exist at temperatures above 161 °C and pressures above several GPa. White tin is metallic, and is the stable crystalline form at or above room temperature. Below 13.2 °C, tin exists in the gray form, which has a diamond cubic crystal structure, similar to diamond, silicon or germanium. Gray tin has no metallic properties at all, is a dull gray powdery material, and has few uses, other than a few specialized semiconductor applications. Although the α–β transformation temperature of tin is nominally 13.2 °C, impurities (e.g. Al, Zn, etc.) lower the transition temperature well below 0 °C, and upon addition of Sb or Bi the transformation may not occur at all.

Physical Properties

Twenty of the 32 crystal classes are piezoelectric, and crystals belonging to one of these classes (point groups) display piezoelectricity. All piezoelectric classes lack a center of symmetry. Any material develops a dielectric polarization when an electric field is applied, but a substance that has such a natural charge separation even in the absence of a field is called a polar material. Whether or not a material is polar is determined solely by its crystal structure. Only 10 of the 32 point groups are polar. All polar crystals are pyroelectric, so the 10 polar crystal classes are sometimes referred to as the pyroelectric classes.

There are a few crystal structures, notably the perovskite structure, which exhibit ferroelectric behavior. This is analogous to ferromagnetism, in that, in the absence of an electric field during production, the ferroelectric crystal does not exhibit a polarization. Upon the application of an electric field of sufficient magnitude, the crystal becomes permanently polarized. This polarization can be reversed by a sufficiently large counter-charge, in the same way that a ferromagnet can be reversed. However, although they are called ferroelectrics, the effect is due to the crystal structure (not the presence of a ferrous metal).

Miller Index

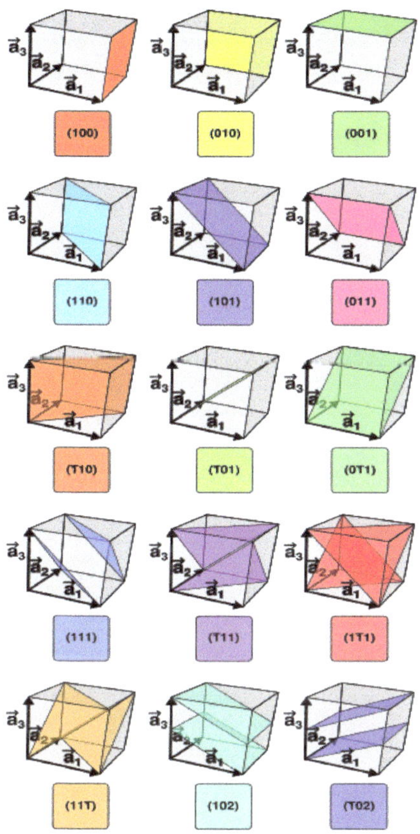

Planes with different Miller indices in cubic crystals

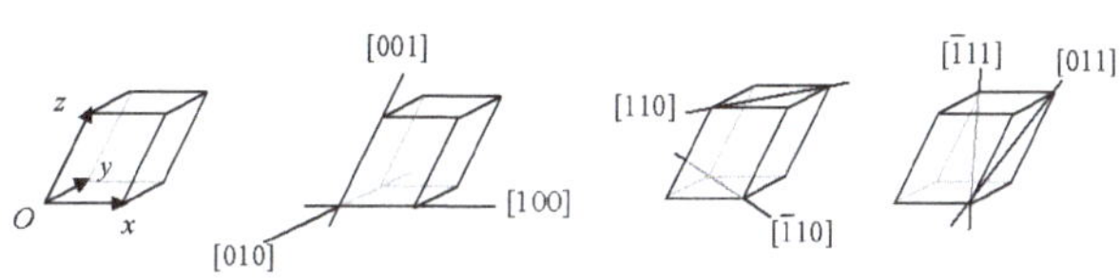

Examples of directions

Miller indices form a notation system in crystallography for planes in crystal (Bravais) lattices.

In particular, a family of lattice planes is determined by three integers h, k, and ℓ, the Miller indices. They are written (hkℓ), and denote the family of planes orthogonal to $h\mathbf{b}_1 + k\mathbf{b}_2 + \ell\mathbf{b}_3$, where $\mathbf{b}_i$ are the basis of the reciprocal lattice vectors. (Note that the plane is not always orthogonal to the linear combination of direct lattice vectors $h\mathbf{a}_1 + k\mathbf{a}_2 + \ell\mathbf{a}_3$ because the reciprocal lattice vectors need not be mutually orthogonal.) By convention, negative integers are written with a bar, as in $\bar{3}$ for −3. The integers are usually written in lowest terms, i.e. their greatest common divisor should be 1.

There are also several related notations:

- the notation {hkℓ} denotes the set of all planes that are equivalent to (hkℓ) by the symmetry of the lattice.

In the context of crystal directions (not planes), the corresponding notations are:

- [hkℓ], with square instead of round brackets, denotes a direction in the basis of the *direct* lattice vectors instead of the reciprocal lattice; and
- similarly, the notation <hkℓ> denotes the set of all directions that are equivalent to [hkℓ] by symmetry.

Miller indices were introduced in 1839 by the British mineralogist William Hallowes Miller. The method was also historically known as the Millerian system, and the indices as Millerian, although this is now rare.

The Miller indices are defined with respect to any choice of unit cell and not only with respect to primitive basis vectors, as is sometimes stated.

Definition

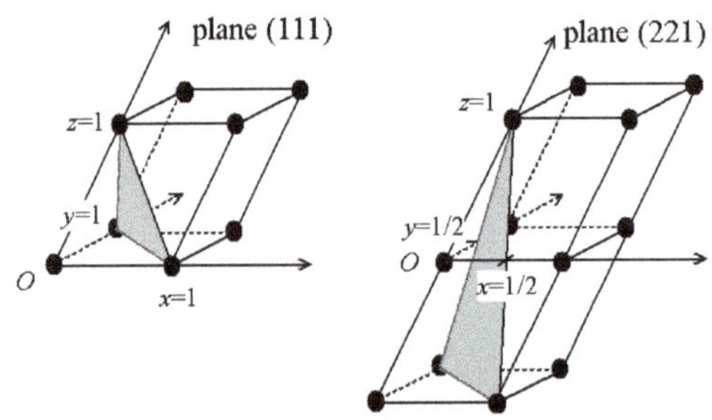

Examples of determining indices for a plane using intercepts with axes; left (111), right (221)

There are two equivalent ways to define the meaning of the Miller indices: via a point in the reciprocal lattice, or as the inverse intercepts along the lattice vectors. Both definitions are given below. In either case, one needs to choose the three lattice vectors a_1, a_2, and a_3 that define the unit cell (note that the conventional unit cell may be larger than the primitive cell of the Bravais lattice, as the examples below illustrate). Given these, the three primitive reciprocal lattice vectors are also determined (denoted b_1, b_2, and b_3).

Then, given the three Miller indices h, k, ℓ, (hkℓ) denotes planes orthogonal to the reciprocal lattice vector:

$$\mathbf{g}_{hk\ell} = h\mathbf{b}_1 + k\mathbf{b}_2 + \ell\mathbf{b}_3.$$

That is, (hkℓ) simply indicates a normal to the planes in the basis of the primitive reciprocal lattice vectors. Because the coordinates are integers, this normal is itself always a reciprocal lattice vector. The requirement of lowest terms means that it is the *shortest* reciprocal lattice vector in the given direction.

Equivalently, (hkℓ) denotes a plane that intercepts the three points a_1/h, a_2/k, and a_3/ℓ, or some multiple thereof. That is, the Miller indices are proportional to the *inverses* of the intercepts of the plane, in the basis of the lattice vectors. If one of the indices is zero, it means that the planes do not intersect that axis (the intercept is "at infinity").

Considering only (hkℓ) planes intersecting one or more lattice points (the lattice planes), the perpendicular distance d between adjacent lattice planes is related to the (shortest) reciprocal lattice vector orthogonal to the planes by the formula: $d = 2\pi / |\mathbf{g}_{hk\ell}|$.

The related notation [hkℓ] denotes the *direction*:

$$h\mathbf{a}_1 + k\mathbf{a}_2 + \ell\mathbf{a}_3.$$

That is, it uses the direct lattice basis instead of the reciprocal lattice. Note that [hkℓ] is *not* generally normal to the (hkℓ) planes, except in a cubic lattice as described below.

Case of Cubic Structures

For the special case of simple cubic crystals, the lattice vectors are orthogonal and of equal length (usually denoted *a*), as are those of the reciprocal lattice. Thus, in this common case, the Miller indices (hkℓ) and [hkℓ] both simply denote normals/directions in Cartesian coordinates.

For cubic crystals with lattice constant *a*, the spacing *d* between adjacent (hkℓ) lattice planes is (from above)

$$d_{hk\ell} = \frac{a}{\sqrt{h^2 + k^2 + \ell^2}}.$$

Because of the symmetry of cubic crystals, it is possible to change the place and sign of the integers and have equivalent directions and planes:

- Indices in *angle brackets* such as ⟨100⟩ denote a *family* of directions which are equivalent due to symmetry operations, such as , , or the negative of any of those directions.
- Indices in *curly brackets* or *braces* such as {100} denote a family of plane normals which are equivalent due to symmetry operations, much the way angle brackets denote a family of directions.

For face-centered cubic and body-centered cubic lattices, the primitive lattice vectors are not orthogonal. However, in these cases the Miller indices are conventionally defined relative to the lattice vectors of the cubic supercell and hence are again simply the Cartesian directions.

Case of Hexagonal and Rhombohedral Structures

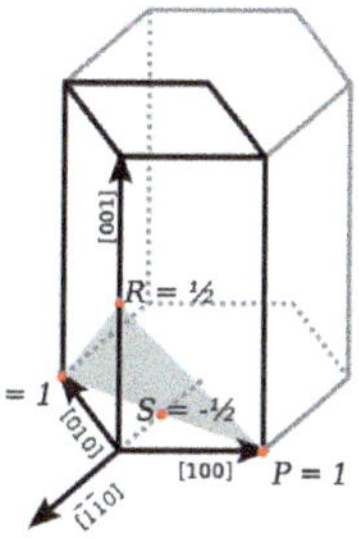

Miller-Bravais indices

With hexagonal and rhombohedral lattice systems, it is possible to use the Bravais-Miller system, which uses four indices ($h\ k\ i\ \ell$) that obey the constraint

$$h + k + i = 0.$$

Here h, k and ℓ are identical to the corresponding Miller indices, and i is a redundant index.

This four-index scheme for labeling planes in a hexagonal lattice makes permutation symmetries apparent. For example, the similarity between (110) ≡ (1120) and (120) ≡ (1210) is more obvious when the redundant index is shown.

In the figure, the (001) plane has a 3-fold symmetry: it remains unchanged by a rotation of 1/3 ($2\pi/3$ rad, 120°). The , and the directions are really similar. If S is the intercept of the plane with the axis, then

$$i = 1/S.$$

There are also *ad hoc* schemes (e.g. in the transmission electron microscopy literature) for indexing hexagonal *lattice vectors* (rather than reciprocal lattice vectors or planes) with four indices. However they don't operate by similarly adding a redundant index to the regular three-index set.

For example, the reciprocal lattice vector (hkℓ) as suggested above can be written in terms of reciprocal lattice vectors as $h\mathbf{b}_1 + k\mathbf{b}_2 + \ell\mathbf{b}_3$. For hexagonal crystals this may be expressed in terms of direct-lattice basis-vectors a1, a2 and a3 as

$$h\mathbf{b}_1 + k\mathbf{b}_2 + \ell\mathbf{b}_3 = \frac{2}{3a^2}(2h+k)\mathbf{a}_1 + \frac{2}{3a^2}(h+2k)\mathbf{a}_2 + \frac{1}{c^2}(\ell)\mathbf{a}_3.$$

Hence zone indices of the direction perpendicular to plane (hkℓ) are, in suitably normalized triplet form, simply $[2h+k, h+2k, \ell(3/2)(a/c)^2]$. When four indices are used for the zone normal to plane (hkℓ), however, the literature often uses $[h, k, -h-k, \ell(3/2)(a/c)^2]$ instead. Thus as you can see, four-index zone indices in square or angle brackets sometimes mix a single direct-lattice index on the right with reciprocal-lattice indices (normally in round or curly brackets) on the left.

Crystallographic Planes and Directions

The crystallographic directions are fictitious lines linking nodes (atoms, ions or molecules) of a crystal. Similarly, the crystallographic planes are fictitious *planes* linking nodes. Some directions and planes have a higher density of nodes; these dense planes have an influence on the behaviour of the crystal:

- optical properties: in condensed matter, the light "jumps" from one atom to the other with the Rayleigh scattering; the velocity of light thus varies according to the directions, whether the atoms are close or far; this gives the birefringence
- adsorption and reactivity: the adsorption and the chemical reactions occur on atoms or molecules, these phenomena are thus sensitive to the density of nodes
- surface tension: the condensation of a material means that the atoms, ions or molecules are more stable if they are surrounded by other similar species; the surface tension of an interface thus varies according to the density on the surface

   - the pores and crystallites tend to have straight grain boundaries following dense planes
   - cleavage
- dislocations (plastic deformation)
   - the dislocation core tends to spread on dense planes (the elastic perturbation is "diluted"); this reduces the friction (Peierls–Nabarro force), the sliding occurs more frequently on dense planes;
   - the perturbation carried by the dislocation (Burgers vector) is along a dense direction: the shift of one node in a dense direction is a lesser distortion;
   - the dislocation line tends to follow a dense direction, the dislocation line is often a straight line, a dislocation loop is often a polygon.

For all these reasons, it is important to determine the planes and thus to have a notation system.

Integer vs. Irrational Miller Indices: Lattice Planes and Quasicrystals

Ordinarily, Miller indices are always integers by definition, and this constraint is physically significant. To understand this, suppose that we allow a plane (abc) where the Miller "indices" a, b and c (defined as above) are not necessarily integers.

If a, b and c have rational ratios, then the same family of planes can be written in terms of integer indices (hkℓ) by scaling a, b and c appropriately: divide by the largest of the three numbers, and then multiply by the least common denominator. Thus, integer Miller indices implicitly include indices with all rational ratios. The reason why planes where the components (in the reciprocal-lattice basis) have rational ratios are of special interest is that these are the lattice planes: they are the only planes whose intersections with the crystal are 2d-periodic.

For a plane (abc) where a, b and c have irrational ratios, on the other hand, the intersection of the plane with the crystal is *not* periodic. It forms an aperiodic pattern known as a quasicrystal. This construction corresponds precisely to the standard "cut-and-project" method of defining a quasicrystal, using a plane with irrational-ratio Miller indices. (Although many quasicrystals, such as the Penrose tiling, are formed by "cuts" of periodic lattices in more than three dimensions, involving the intersection of more than one such hyperplane.)

X-ray Crystallography

X-ray crystallography is a technique used for determining the atomic and molecular structure of a crystal, in which the crystalline atoms cause a beam of incident X-rays to diffract into many specific directions. By measuring the angles and intensities of these diffracted beams, a crystallographer can produce a three-dimensional picture of the density of electrons within the crystal. From this electron density, the mean positions of the atoms in the crystal can be determined, as well as their chemical bonds, their disorder, and various other information.

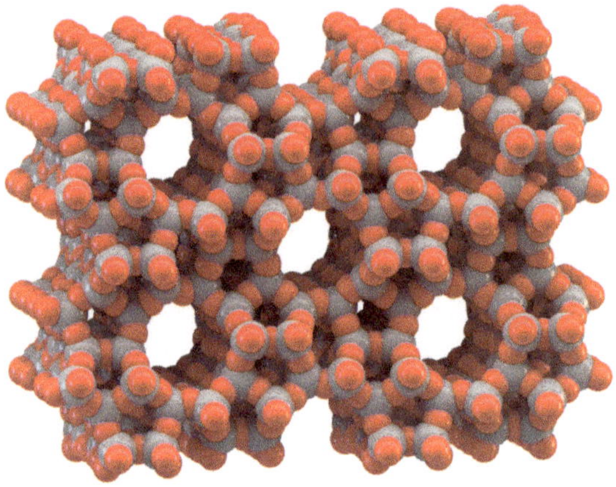

X-ray crystallography provides an atomistic model of zeolite, an aluminosilicate.

Since many materials can form crystals—such as salts, metals, minerals, semiconductors, as well as various inorganic, organic, and biological molecules—X-ray crystallography has been fundamental in the development of many scientific fields. In its first decades of use, this method determined the size of atoms, the lengths and types of chemical bonds, and the atomic-scale differences among various materials, especially minerals and alloys. The method also revealed the structure and function of many biological molecules, including vitamins, drugs, proteins and nucleic acids such as DNA. X-ray crystallography is still the chief method for characterizing the atomic structure of new materials and in discerning materials that appear similar by other experiments. X-ray crystal structures can also account for unusual electronic or elastic properties of a material, shed light on chemical interactions and processes, or serve as the basis for designing pharmaceuticals against diseases.

In a single-crystal X-ray diffraction measurement, a crystal is mounted on a goniometer. The goniometer is used to position the crystal at selected orientations. The crystal is illuminated with a finely focused monochromatic beam of X-rays, producing a diffraction pattern of regularly spaced spots known as *reflections*. The two-dimensional images taken at different orientations are converted into a three-dimensional model of the density of electrons within the crystal using the mathematical method of Fourier transforms, combined with chemical data known for the sample. Poor resolution (fuzziness) or even errors may result if the crystals are too small, or not uniform enough in their internal makeup.

X-ray crystallography is related to several other methods for determining atomic structures. Similar diffraction patterns can be produced by scattering electrons or neutrons, which are likewise interpreted by Fourier transformation. If single crystals of sufficient size cannot be obtained, various other X-ray methods can be applied to obtain less detailed information; such methods include fiber diffraction, powder diffraction and (if the sample is not crystallized) small-angle X-ray scattering (SAXS). If the material under investigation is only available in the form of nanocrystalline powders or suffers from poor crystallinity, the methods of electron crystallography can be applied for determining the atomic structure.

For all above mentioned X-ray diffraction methods, the scattering is elastic; the scattered X-rays have the same wavelength as the incoming X-ray. By contrast, *inelastic* X-ray scattering methods are useful in studying excitations of the sample, rather than the distribution of its atoms.

History

Early Scientific History of Crystals and X-rays

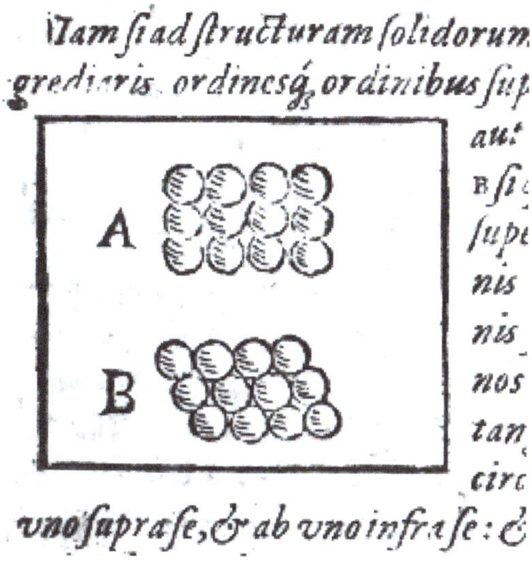

Drawing of square (Figure A, above) and hexagonal (Figure B, below) packing from Kepler's work, *Strena seu de Nive Sexangula.*

Crystals, though long admired for their regularity and symmetry, were not investigated scientifically until the 17th century. Johannes Kepler hypothesized in his work *Strena seu de Nive Sexangula* (A New Year's Gift of Hexagonal Snow) (1611) that the hexagonal symmetry of snowflake crystals was due to a regular packing of spherical water particles.

As shown by X-ray crystallography, the hexagonal symmetry of snowflakes results from the tetrahedral arrangement of hydrogen bonds about each water molecule. The water molecules are arranged similarly to the silicon atoms in the tridymite polymorph of SiO_2. The resulting crystal structure has hexagonal symmetry when viewed along a principal axis.

The Danish scientist Nicolas Steno (1669) pioneered experimental investigations of crystal symmetry. Steno showed that the angles between the faces are the same in every exemplar of a particular type of crystal, and René Just Haüy (1784) discovered that every face of a crystal can be described by simple stacking patterns of blocks of the same shape and size. Hence, William Hallowes Miller in 1839 was able to give each face a unique label of three small integers, the Miller indices which remain in use today for identifying crystal faces. Haüy's study led to the correct idea that crystals are a regular three-dimensional array (a Bravais lattice) of atoms and molecules; a single unit cell is repeated indefinitely along three principal directions that are not necessarily perpendicular. In the 19th century, a complete catalog of the possible symmetries of a crystal was worked out by Johan Hessel, Auguste Bravais, Evgraf Fedorov, Arthur Schönflies and (belatedly) William Barlow (1894). From the available data and physical reasoning, Barlow proposed several crystal structures in the 1880s that were validated later by X-ray crystallography; however, the available data were too scarce in the 1880s to accept his models as conclusive.

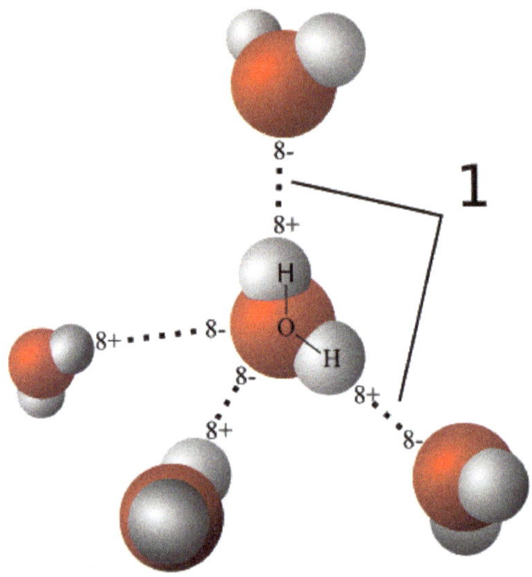

X-ray crystallography shows the arrangement of water molecules in ice, revealing the hydrogen bonds (1) that hold the solid together. Few other methods can determine the structure of matter with such precision (*resolution*).

Wilhelm Röntgen discovered X-rays in 1895, just as the studies of crystal symmetry were being concluded. Physicists were initially uncertain of the nature of X-rays, but soon suspected (correctly) that they were waves of electromagnetic radiation, in other words, another form of light. At that time, the wave model of light—specifically, the Maxwell theory of electromagnetic radiation—was well accepted among scientists, and experiments by Charles Glover Barkla showed that X-rays exhibited phenomena associated with electromagnetic waves, including transverse polarization and spectral lines akin to those observed in the visible wavelengths. Single-slit experiments in the laboratory of Arnold Sommerfeld suggested that X-rays had a wavelength of about 1 angstrom. However, X-rays are composed of photons, and thus are not only waves of electromagnetic radiation but also exhibit particle-like properties. Albert Einstein introduced the photon concept in 1905, but it was not broadly accepted until 1922, when Arthur Compton confirmed it by the scattering of X-rays from electrons. Therefore, these particle-like properties of X-rays, such as their ionization of gases, caused William Henry Bragg to argue in 1907 that X-rays were *not* electromagnetic radiation. Nevertheless, Bragg's view was not broadly accepted and the observation of X-ray diffraction by Max von Laue in 1912 confirmed for most scientists that X-rays were a form of electromagnetic radiation.

X-ray Diffraction

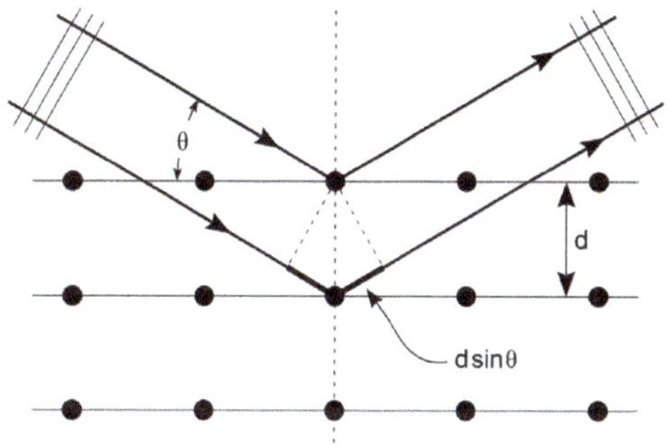

The incoming beam (coming from upper left) causes each scatterer to re-radiate a small portion of its intensity as a spherical wave. If scatterers are arranged symmetrically with a separation *d*, these spherical waves will be in sync (add constructively) only in directions where their path-length difference $2d \sin \theta$ equals an integer multiple of the wavelength λ. In that case, part of the incoming beam is deflected by an angle 2θ, producing a *reflection* spot in the diffraction pattern.

Crystals are regular arrays of atoms, and X-rays can be considered waves of electromagnetic radiation. Atoms scatter X-ray waves, primarily through the atoms' electrons. Just as an ocean wave striking a lighthouse produces secondary circular waves emanating from the lighthouse, so an X-ray striking an electron produces secondary spherical waves emanating from the electron. This phenomenon is known as elastic scattering, and the electron (or lighthouse) is known as the *scatterer*. A regular array of scatterers produces a regular array of spherical waves. Although these waves cancel one another out in most directions through destructive interference, they add constructively in a few specific directions, determined by Bragg's law:

$$2d \sin \theta = n\lambda$$

Here d is the spacing between diffracting planes, θ is the incident angle, n is any integer, and λ is the wavelength of the beam. These specific directions appear as spots on the diffraction pattern called reflections. Thus, X-ray diffraction results from an electromagnetic wave (the X-ray) impinging on a regular array of scatterers (the repeating arrangement of atoms within the crystal).

X-rays are used to produce the diffraction pattern because their wavelength λ is typically the same order of magnitude (1–100 angstroms) as the spacing *d* between planes in the crystal. In principle, any wave impinging on a regular array of scatterers produces diffraction, as predicted first by Francesco Maria Grimaldi in 1665. To produce significant diffraction, the spacing between the scatterers and the wavelength of the impinging wave should be similar in size. For illustration, the diffraction of sunlight through a bird's feather was first reported by James Gregory in the later 17th century. The first artificial diffraction gratings for visible light were constructed by David Rittenhouse in 1787, and Joseph von Fraunhofer in 1821. However, visible light has too long a wavelength (typically, 5500 angstroms) to observe diffraction from crystals. Prior to the first X-ray diffraction experiments, the spacings between lattice planes in a crystal were not known with certainty.

The idea that crystals could be used as a diffraction grating for X-rays arose in 1912 in a conversation between Paul Peter Ewald and Max von Laue in the English Garden in Munich. Ewald had proposed a resonator model of crystals for his thesis, but this model could not be validated using visible light, since the wavelength was much larger than the spacing between the resonators. Von Laue realized that electromagnetic radiation of a shorter wavelength was needed to observe such small spacings, and suggested that X-rays might have a wavelength comparable to the unit-cell spacing in crystals. Von Laue worked with two technicians, Walter Friedrich and his assistant Paul Knipping, to shine a beam of X-rays through a copper sulfate crystal and record its diffraction on a photographic plate. After being developed, the plate showed a large number of well-defined spots arranged in a pattern of intersecting circles around the spot produced by the central beam. Von Laue developed a law that connects the scattering angles and the size and orientation of the unit-cell spacings in the crystal, for which he was awarded the Nobel Prize in Physics in 1914.

Scattering

As described in the mathematical derivation below, the X-ray scattering is determined by the density of electrons within the crystal. Since the energy of an X-ray is much greater than that of a valence electron, the scattering may be modeled as Thomson scattering, the interaction of an electromagnetic ray with a free electron. This model is generally adopted to describe the polarization of the scattered radiation.

The intensity of Thomson scattering for one particle with mass m and charge q is:

$$I_o = I_e\left(\frac{q^4}{m^2c^4}\right)\frac{1+\cos^2 2\theta}{2} = I_e 7.94.10^{-26}\frac{1+\cos^2 2\theta}{2} = I_e f$$

Hence the atomic nuclei, which are much heavier than an electron, contribute negligibly to the scattered X-rays.

Development from 1912 to 1920

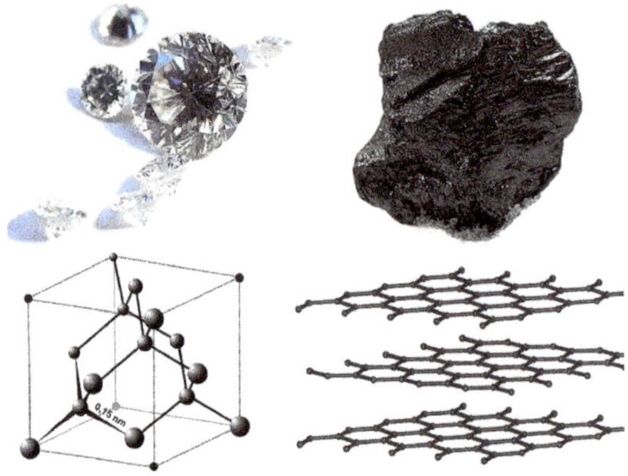

Although diamonds (top left) and graphite (top right) are identical in chemical composition—being both pure carbon—X-ray crystallography revealed the arrangement of their atoms (bottom) accounts for their different properties. In diamond, the carbon atoms are arranged tetrahedrally and held together by single covalent bonds, making it strong in all directions. By contrast, graphite is composed of stacked sheets. Within the sheet, the bonding is covalent and has hexagonal symmetry, but there are no covalent bonds between the sheets, making graphite easy to cleave into flakes.

After Von Laue's pioneering research, the field developed rapidly, most notably by physicists William Lawrence Bragg and his father William Henry Bragg. In 1912–1913, the younger Bragg developed Bragg's law, which connects the observed scattering with reflections from evenly spaced planes within the crystal. The Braggs, father and son, shared the 1915 Nobel Prize in Physics for their work in crystallography. The earliest structures were generally simple and marked by one-dimensional symmetry. However, as computational and experimental methods improved over the next decades, it became feasible to deduce reliable atomic positions for more complicated two- and three-dimensional arrangements of atoms in the unit-cell.

The potential of X-ray crystallography for determining the structure of molecules and minerals—then only known vaguely from chemical and hydrodynamic experiments—was realized immediately. The earliest structures were simple inorganic crystals and minerals, but even these revealed fundamental laws of physics and chemistry. The first atomic-resolution structure to be "solved" (i.e., determined) in 1914 was that of table salt. The distribution of electrons in the table-salt structure showed that crystals are not necessarily composed of covalently bonded molecules, and proved the existence of ionic compounds. The structure of diamond was solved in the same year, proving the tetrahedral arrangement of its chemical bonds and showing that the length of C–C single bond was 1.52 angstroms. Other early structures included copper, calcium fluoride (CaF_2, also known as *fluorite*), calcite ($CaCO_3$) and pyrite (FeS_2) in 1914; spinel ($MgAl_2O_4$) in 1915; the rutile and anatase forms of titanium dioxide (TiO_2) in 1916; pyrochroite $Mn(OH)_2$ and, by extension, brucite $Mg(OH)_2$ in 1919;. Also in 1919 sodium nitrate ($NaNO_3$) and caesium dichloroiodide ($CsICl_2$) were determined by Ralph Walter Graystone Wyckoff, and the wurtzite (hexagonal ZnS) structure became known in 1920.

The structure of graphite was solved in 1916 by the related method of powder diffraction, which was developed by Peter Debye and Paul Scherrer and, independently, by Albert Hull in 1917. The structure of graphite was determined from single-crystal diffraction in 1924 by two groups independently. Hull also used the powder method to determine the structures of various metals, such as iron and magnesium.

Cultural and Aesthetic Importance

In what has been called his scientific autobiography, *The Development of X-ray Analysis*, Sir William Lawrence Bragg mentioned that he believed the field of crystallography was particularly welcoming to women because the techno-aesthetics of the molecular structures resembled textiles and household objects. Bragg was known to compare crystal formation to "curtains, wallpapers, mosaics, and roses".

In 1951, the Festival Pattern Group at the Festival of Britain hosted a collaborative group of textile manufacturers and experienced crystallographers to design lace and prints based on the X-ray crystallography of insulin, china clay, and hemoglobin. One of the leading scientists of the project was Dr. Helen Megaw (1907–2002), the Assistant Director of Research at the Cavendish Laboratory in Cambridge at the time. Megaw is credited as one of the central figures who took inspiration from crystal diagrams and saw their potential in design. In 2008, the Wellcome Collection in London curated an exhibition on the Festival Pattern Group called "From Atom to Patterns".

Contributions to Chemistry and Material Science

X-ray crystallography has led to a better understanding of chemical bonds and non-covalent interactions. The initial studies revealed the typical radii of atoms, and confirmed many theoretical models of chemical bonding, such as the tetrahedral bonding of carbon in the diamond structure, the octahedral bonding of metals observed in ammonium hexachloroplatinate (IV), and the resonance observed in the planar carbonate group and in aromatic molecules. Kathleen Lonsdale's 1928 structure of hexamethylbenzene established the hexagonal symmetry of benzene and showed a clear difference in bond length between the aliphatic C–C bonds and aromatic C–C bonds; this finding led to the idea of resonance between chemical bonds, which had profound consequences for the development of chemistry. Her conclusions were anticipated by William Henry Bragg, who published models of naphthalene and anthracene in 1921 based on other molecules, an early form of molecular replacement.

Also in the 1920s, Victor Moritz Goldschmidt and later Linus Pauling developed rules for eliminating chemically unlikely structures and for determining the relative sizes of atoms. These rules led to the structure of brookite (1928) and an understanding of the relative stability of the rutile, brookite and anatase forms of titanium dioxide.

The distance between two bonded atoms is a sensitive measure of the bond strength and its bond order; thus, X-ray crystallographic studies have led to the discovery of even more exotic types of bonding in inorganic chemistry, such as metal-metal double bonds, metal-metal quadruple bonds, and three-center, two-electron bonds. X-ray crystallography—or, strictly speaking, an inelastic Compton scattering experiment—has also provided evidence for the partly covalent character of hydrogen bonds. In the field of organometallic chemistry, the X-ray structure of ferrocene initiated scientific studies of sandwich compounds, while that of Zeise's salt stimulated research into "back bonding" and metal-pi complexes. Finally, X-ray crystallography had a pioneering role in the development of supramolecular chemistry, particularly in clarifying the structures of the crown ethers and the principles of host-guest chemistry.

In material sciences, many complicated inorganic and organometallic systems have been analyzed using single-crystal methods, such as fullerenes, metalloporphyrins, and other complicated compounds. Single-crystal diffraction is also used in the pharmaceutical industry, due to recent problems with polymorphs. The major factors affecting the quality of single-crystal structures are the crystal's size and regularity; recrystallization is a commonly used technique to improve these factors in small-molecule crystals. The Cambridge Structural Database contains over 800,000 structures as of September 2016; over 99% of these structures were determined by X-ray diffraction.

Mineralogy and Metallurgy

Since the 1920s, X-ray diffraction has been the principal method for determining the arrangement of atoms in minerals and metals. The application of X-ray crystallography to mineralogy began with the structure of garnet, which was determined in 1924 by Menzer. A systematic X-ray crystallographic study of the silicates was undertaken in the 1920s. This study showed that, as the Si/O ratio is altered, the silicate crystals exhibit significant changes in their atomic arrangements. Machatschki extended these insights to minerals in which aluminium substitutes for the silicon atoms of the silicates. The first application of X-ray crystallography to metallurgy likewise occurred in the mid-1920s. Most notably, Linus Pauling's structure of the alloy Mg_2Sn led to his theory of the stability and structure of complex ionic crystals.

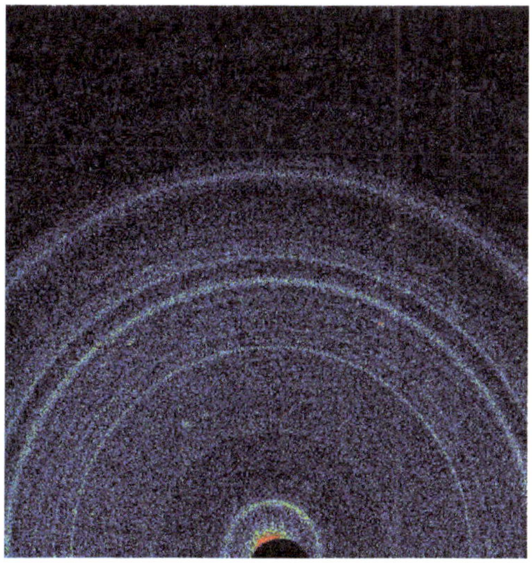

First X-ray diffraction view of Martian soil – CheMin analysis reveals feldspar, pyroxenes, olivine and more (Curiosity rover at "Rocknest", October 17, 2012).

On October 17, 2012, the Curiosity rover on the planet Mars at "Rocknest" performed the first X-ray diffraction analysis of Martian soil. The results from the rover's CheMin analyzer revealed the presence of several minerals, including feldspar, pyroxenes and olivine, and suggested that the Martian soil in the sample was similar to the "weathered basaltic soils" of Hawaiian volcanoes.

Early Organic and Small Biological Molecules

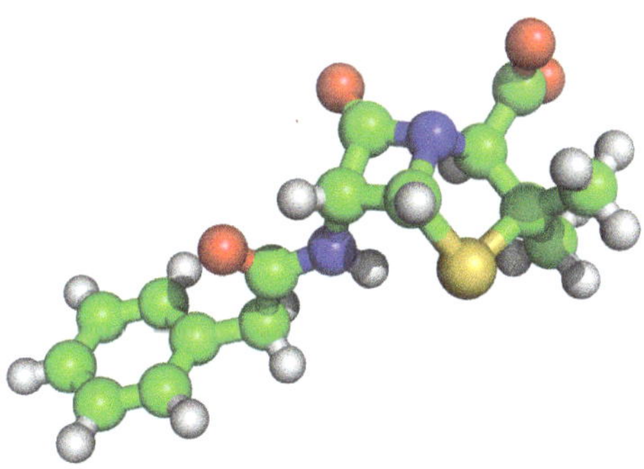

The three-dimensional structure of penicillin, solved by Dorothy Crowfoot Hodgkin in 1945. The green, white, red, yellow and blue spheres represent atoms of carbon, hydrogen, oxygen, sulfur and nitrogen, respectively.

The first structure of an organic compound, hexamethylenetetramine, was solved in 1923. This was followed by several studies of long-chain fatty acids, which are an important component of biological membranes. In the 1930s, the structures of much larger molecules with two-dimensional complexity began to be solved. A significant advance was the structure of phthalocyanine, a large planar molecule that is closely related to porphyrin molecules important in biology, such as heme, corrin and chlorophyll.

X-ray crystallography of biological molecules took off with Dorothy Crowfoot Hodgkin, who solved the structures of cholesterol (1937), penicillin (1946) and vitamin B12 (1956), for which she was awarded the Nobel Prize in Chemistry in 1964. In 1969, she succeeded in solving the structure of insulin, on which she worked for over thirty years.

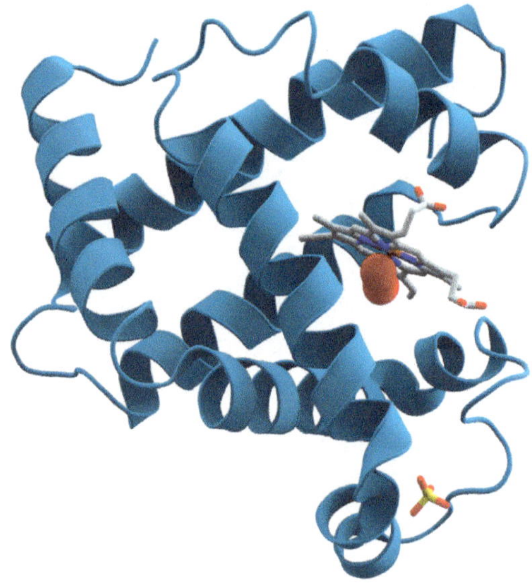

Ribbon diagram of the structure of myoglobin, showing colored alpha helices. Such proteins are long, linear molecules with thousands of atoms; yet the relative position of each atom has been determined with sub-atomic resolution by X-ray crystallography. Since it is difficult to visualize all the atoms at once, the ribbon shows the rough path of the protein polymer from its N-terminus (blue) to its C-terminus (red).

Biological Macromolecular Crystallography

Crystal structures of proteins (which are irregular and hundreds of times larger than cholesterol) began to be solved in the late 1950s, beginning with the structure of sperm whale myoglobin by Sir John Cowdery Kendrew, for which he shared the Nobel Prize in Chemistry with Max Perutz in 1962. Since that success, 132055 X-ray crystal structures of proteins, nucleic acids and other biological molecules have been determined. For comparison, the nearest competing method in terms of structures analyzed is nuclear magnetic resonance (NMR) spectroscopy, which has resolved 11904 chemical structures. Moreover, crystallography can solve structures of arbitrarily large molecules, whereas solution-state NMR is restricted to relatively small ones (less than 70 kDa). X-ray crystallography is now used routinely by scientists to determine how a pharmaceutical drug interacts with its protein target and what changes might improve it. However, intrinsic membrane proteins remain challenging to crystallize because they require detergents or other means to solubilize them in isolation, and such detergents often interfere with crystallization. Such membrane proteins are a large component of the genome, and include many proteins of great physiological importance, such as ion channels and receptors. Helium cryogenics are used to prevent radiation damage in protein crystals.

On the other end of the size scale, even relatively small molecules may pose challenges for the resolving power of X-ray crystallography. The structure assigned in 1991 to the antibiotic isolated from a marine organism, diazonamide A ($C_{40}H_{34}Cl_2N_6O_6$, molar mass 765.65 g/mol), proved to be incorrect by the classical proof of structure: a synthetic sample was not identical to the natural product. The mistake was attributed to the inability of X-ray crystallography to distinguish between the correct -OH / >NH and the interchanged -NH_2 / -O- groups in the incorrect structure. With advances in instrumentation, however, it is now routinely possible to distinguish between such similar groups using modern single-crystal X-ray diffractometers.

Relationship to Other Scattering Techniques

Elastic vs. Inelastic Scattering

X-ray crystallography is a form of elastic scattering; the outgoing X-rays have the same energy, and thus same wavelength, as the incoming X-rays, only with altered direction. By contrast, *inelastic scattering* occurs when energy is transferred from the incoming X-ray to the crystal, e.g., by exciting an inner-shell electron to a higher energy level. Such inelastic scattering reduces the energy (or increases the wavelength) of the outgoing beam. Inelastic scattering is useful for probing such excitations of matter, but not in determining the distribution of scatterers within the matter, which is the goal of X-ray crystallography.

X-rays range in wavelength from 10 to 0.01 nanometers; a typical wavelength used for crystallography is 1 Å (0.1 nm), which is on the scale of covalent chemical bonds and the radius of a single atom. Longer-wavelength photons (such as ultraviolet radiation) would not have sufficient resolution to determine the atomic positions. At the other extreme, shorter-wavelength photons such as gamma rays are difficult to produce in large numbers, difficult to focus, and interact too strongly with matter, producing particle-antiparticle pairs. Therefore, X-rays are the "sweetspot" for wavelength when determining atomic-resolution structures from the scattering of electromagnetic radiation.

Other X-ray Techniques

Other forms of elastic X-ray scattering include powder diffraction, Small-Angle X-ray Scattering (SAXS) and several types of X-ray fiber diffraction, which was used by Rosalind Franklin in determining the double-helix structure of DNA. In general, single-crystal X-ray diffraction offers more structural information than these other techniques; however, it requires a sufficiently large and regular crystal, which is not always available.

These scattering methods generally use *monochromatic* X-rays, which are restricted to a single wavelength with minor deviations. A broad spectrum of X-rays (that is, a blend of X-rays with different wavelengths) can also be used to carry out X-ray diffraction, a technique known as the Laue method. This is the method used in the original discovery of X-ray diffraction. Laue scattering provides much structural information with only a short exposure to the X-ray beam, and is therefore used in structural studies of very rapid events (Time resolved crystallography). However, it is not as well-suited as monochromatic scattering for determining the full atomic structure of a crystal and therefore works better with crystals with relatively simple atomic arrangements.

The Laue back reflection mode records X-rays scattered backwards from a broad spectrum source. This is useful if the sample is too thick for X-rays to transmit through it. The diffracting planes in the crystal are determined by knowing that the normal to the diffracting plane bisects the angle between the incident beam and the diffracted beam. A Greninger chart can be used to interpret the back reflection Laue photograph.

Electron and Neutron Diffraction

Other particles, such as electrons and neutrons, may be used to produce a diffraction pattern. Although electron, neutron, and X-ray scattering are based on different physical processes, the resulting diffraction patterns are analyzed using the same coherent diffraction imaging techniques.

As derived below, the electron density within the crystal and the diffraction patterns are related by a simple mathematical method, the Fourier transform, which allows the density to be calculated relatively easily from the patterns. However, this works only if the scattering is *weak*, i.e., if the scattered beams are much less intense than the incoming beam. Weakly scattered beams pass through the remainder of the crystal without undergoing a second scattering event. Such re-scattered waves are called "secondary scattering" and hinder the analysis. Any sufficiently thick crystal will produce secondary scattering, but since X-rays interact relatively weakly with the electrons, this is generally not a significant concern. By contrast, electron beams may produce strong secondary scattering even for relatively thin crystals (>100 nm). Since this thickness corresponds to the diameter of many viruses, a promising direction is the electron diffraction of isolated macromolecular assemblies, such as viral capsids and molecular machines, which may be carried out with a cryo-electron microscope. Moreover, the strong interaction of electrons with matter (about 1000 times stronger than for X-rays) allows determination of the atomic structure of extremely small volumes. The field of applications for electron crystallography ranges from bio molecules like membrane proteins over organic thin films to the complex structures of (nanocrystalline) intermetallic compounds and zeolites.

Neutron diffraction is an excellent method for structure determination, although it has been difficult to obtain intense, monochromatic beams of neutrons in sufficient quantities. Traditionally, nuclear reactors have been used, although sources producing neutrons by spallation are becoming increasingly available. Being uncharged, neutrons scatter much more readily from the atomic nuclei rather than from the electrons. Therefore, neutron scattering is very useful for observing the positions of light atoms with few electrons, especially hydrogen, which is essentially invisible in the X-ray diffraction. Neutron scattering also has the remarkable property that the solvent can be made invisible by adjusting the ratio of normal water, H_2O, and heavy water, D_2O.

Methods

Overview of Single-crystal X-ray Diffraction

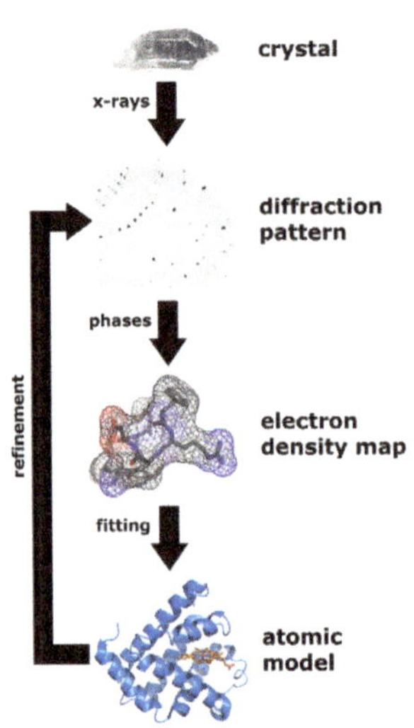

Workflow for solving the structure of a molecule by X-ray crystallography

The oldest and most precise method of X-ray crystallography is *single-crystal X-ray diffraction*, in which a beam of X-rays strikes a single crystal, producing scattered beams. When they land on a piece of film or other detector, these beams make a *diffraction pattern* of spots; the strengths and angles of these beams are recorded as the crystal is gradually rotated. Each spot is called a *reflection*, since it corresponds to the reflection of the X-rays from one set of evenly spaced planes within the crystal. For single crystals of sufficient purity and regularity, X-ray diffraction data can determine the mean chemical bond lengths and angles to within a few thousandths of an angstrom and to within a few tenths of a degree, respectively. The atoms in a crystal are not static, but oscillate about their mean positions, usually by less than a few tenths of an angstrom. X-ray crystallography allows measuring the size of these oscillations.

Procedure

The technique of single-crystal X-ray crystallography has three basic steps. The first—and often most difficult—step is to obtain an adequate crystal of the material under study. The crystal should be sufficiently large (typically larger than 0.1 mm in all dimensions), pure in composition and regular in structure, with no significant internal imperfections such as cracks or twinning.

In the second step, the crystal is placed in an intense beam of X-rays, usually of a single wavelength (*monochromatic X-rays*), producing the regular pattern of reflections. As the crystal is gradually rotated, previous reflections disappear and new ones appear; the intensity of every spot is recorded at every orientation of the crystal. Multiple data sets may have to be collected, with each set covering slightly more than half a full rotation of the crystal and typically containing tens of thousands of reflections.

In the third step, these data are combined computationally with complementary chemical information to produce and refine a model of the arrangement of atoms within the crystal. The final, refined model of the atomic arrangement—now called a *crystal structure*—is usually stored in a public database.

Limitations

As the crystal's repeating unit, its unit cell, becomes larger and more complex, the atomic-level picture provided by X-ray crystallography becomes less well-resolved (more "fuzzy") for a given number of observed reflections. Two limiting cases of X-ray crystallography—"small-molecule" (which includes continuous inorganic solids) and "macromolecular" crystallography—are often discerned. *Small-molecule crystallography* typically involves crystals with fewer than 100 atoms in their asymmetric unit; such crystal structures are usually so well resolved that the atoms can be discerned as isolated "blobs" of electron density. By contrast, *macromolecular crystallography* often involves tens of thousands of atoms in the unit cell. Such crystal structures are generally less well-resolved (more "smeared out"); the atoms and chemical bonds appear as tubes of electron density, rather than as isolated atoms. In general, small molecules are also easier to crystallize than macromolecules; however, X-ray crystallography has proven possible even for viruses with hundreds of thousands of atoms. Though normally x-ray crystallography can only be performed if the sample is in crystal form, new research has been done into sampling non-crystalline forms of samples.

Crystallization

A protein crystal seen under a microscope. Crystals used in X-ray crystallography may be smaller than a millimeter across.

Although crystallography can be used to characterize the disorder in an impure or irregular crystal, crystallography generally requires a pure crystal of high regularity to solve the structure of a complicated arrangement of atoms. Pure, regular crystals can sometimes be obtained from natural or synthetic materials, such as samples of metals, minerals or other macroscopic materials. The regularity of such crystals can sometimes be improved with macromolecular crystal annealing and other methods. However, in many cases, obtaining a diffraction-quality crystal is the chief barrier to solving its atomic-resolution structure.

Small-molecule and macromolecular crystallography differ in the range of possible techniques used to produce diffraction-quality crystals. Small molecules generally have few degrees of conformational freedom, and may be crystallized by a wide range of methods, such as chemical vapor deposition and recrystallization. By contrast, macromolecules generally have many degrees of freedom and their crystallization must be carried out so as to maintain a stable structure. For example, proteins and larger RNA molecules cannot be crystallized if their tertiary structure has been unfolded; therefore, the range of crystallization conditions is restricted to solution conditions in which such molecules remain folded.

Protein crystals are almost always grown in solution. The most common approach is to lower the solubility of its component molecules very gradually; if this is done too quickly, the molecules will precipitate from solution, forming a useless dust or amorphous gel on the bottom of the container. Crystal growth in solution is characterized by two steps: *nucleation* of a microscopic crystallite (possibly having only 100 molecules), followed by *growth* of that crystallite, ideally to a diffraction-quality crystal. The solution conditions that favor the first step (nucleation) are not always the same conditions that favor the second step (subsequent growth). The crystallographer's goal is to identify solution conditions that favor the development of a single, large crystal, since larger crystals offer improved resolution of the molecule. Consequently, the solution conditions should *disfavor* the first step (nucleation) but *favor* the second (growth), so that only one large crystal forms per droplet. If nucleation is favored too much, a shower of small crystallites will form in the droplet, rather than one large crystal; if favored too little, no crystal will form whatsoever. Other approaches involves, crystallizing proteins under oil, where aqueous protein solutions are dispensed under liquid oil, and water evaporates through the layer of oil. Different oils have different evaporation permeabilities, therefore yielding changes in concentration rates from different percipient/protein mixture. The technique relies on bringing the protein directly into the nucleation zone by mixing protein with the appropriate amount of percipient to prevent the diffusion of water out of the drop.

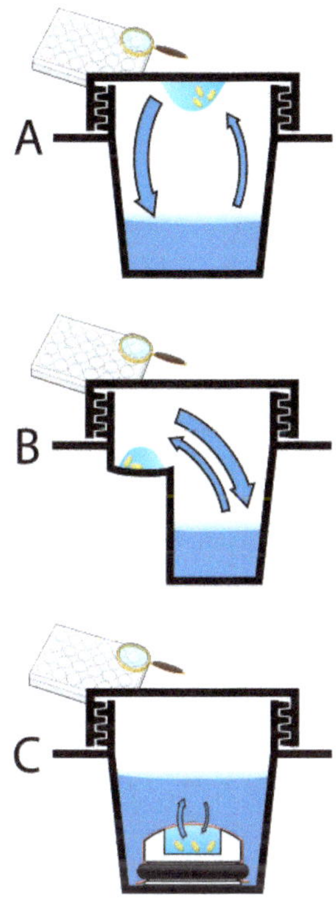

Three methods of preparing crystals, A: Hanging drop. B: Sitting drop. C: Microdialysis

It is extremely difficult to predict good conditions for nucleation or growth of well-ordered crystals. In practice, favorable conditions are identified by *screening*; a very large batch of the molecules is prepared, and a wide variety of crystallization solutions are tested. Hundreds, even thousands, of solution conditions are generally tried before finding the successful one. The various conditions can use one or more physical mechanisms to lower the solubility of the molecule; for example, some may change the pH, some contain salts of the Hofmeister series or chemicals that lower the dielectric constant of the solution, and still others contain large polymers such as polyethylene glycol that drive the molecule out of solution by entropic effects. It is also common to try several temperatures for encouraging crystallization, or to gradually lower the temperature so that the solution becomes supersaturated. These methods require large amounts of the target molecule, as they use high concentration of the molecule(s) to be crystallized. Due to the difficulty in obtaining such large quantities (milligrams) of crystallization-grade protein, robots have been developed that are capable of accurately dispensing crystallization trial drops that are in the order of 100 nanoliters in volume. This means that 10-fold less protein is used per experiment when compared to crystallization trials set up by hand (in the order of 1 microliter).

Several factors are known to inhibit or mar crystallization. The growing crystals are generally held at a constant temperature and protected from shocks or vibrations that might disturb their crystallization. Impurities in the molecules or in the crystallization solutions are often inimical to crystallization. Conformational flexibility in the molecule also tends to make crystallization less likely, due to entropy. Ironically, molecules that tend to self-assemble into regular helices are often unwilling to assemble into crystals. Crystals can be marred by twinning, which can occur when a unit cell can pack equally favorably in multiple orientations; although recent advances in computational methods may allow solving the structure of some twinned crystals. Having failed to crystallize a target molecule, a crystallographer may try again with a slightly modified version of the molecule; even small changes in molecular properties can lead to large differences in crystallization behavior.

Data Collection

Mounting the Crystal

The crystal is mounted for measurements so that it may be held in the X-ray beam and rotated. There are several methods of mounting. In the past, crystals were loaded into glass capillaries with the crystallization solution (the mother liquor). Nowadays, crystals of small molecules are typically attached with oil or glue to a glass fiber or a loop, which is made of nylon or plastic and attached to a solid rod. Protein crystals are scooped up by a loop, then flash-frozen with liquid nitrogen. This freezing reduces the radiation damage of the X-rays, as well as the noise in the Bragg peaks due to thermal motion (the Debye-Waller effect). However, untreated protein crystals often crack if flash-frozen; therefore, they are generally pre-soaked in a cryoprotectant solution before freezing. Unfortunately, this pre-soak may itself cause the crystal to crack, ruining it for crystallography. Generally, successful cryo-conditions are identified by trial and error.

The capillary or loop is mounted on a goniometer, which allows it to be positioned accurately within the X-ray beam and rotated. Since both the crystal and the beam are often very small, the crystal must be centered within the beam to within ~25 micrometers accuracy, which is aided by a camera focused on the crystal. The most common type of goniometer is the "kappa goniometer", which offers three angles of rotation: the ω angle, which rotates about an axis perpendicular to the beam; the κ angle, about an axis at ~50° to the ω axis; and, finally, the φ angle about the loop/capillary axis. When the κ angle is zero, the ω and φ axes are aligned. The κ rotation allows for convenient mounting of the crystal, since the arm in which the crystal is mounted may be swung out towards the crystallographer. The oscillations carried out during data collection (mentioned below) involve the ω axis only. An older type of goniometer is the four-circle goniometer, and its relatives such as the six-circle goniometer.

X-ray Sources

Rotating Anode

Small scale can be done on a local X-ray tube source, typically coupled with an image plate detector. These have the advantage of being (relatively) inexpensive and easy to maintain, and allow for quick screening and collection of samples. However, the wavelength light produced is limited by anode material, typically copper. Further, intensity is limited by the power applied and cooling capacity available to avoid melting the anode. In such systems, electrons are boiled off of a cathode and accelerated through a strong electric potential of ~50 kV; having reached a high speed, the electrons collide with a metal plate, emitting *bremsstrahlung* and some strong spectral lines corresponding to the excitation of inner-shell electrons of the metal. The most common metal used is copper, which can be kept cool easily, due to its high thermal conductivity, and which produces strong K_α and K_β lines. The K_β line is sometimes suppressed with a thin (~10 μm) nickel foil. The simplest and cheapest variety of sealed X-ray tube has a stationary anode (the Crookes tube) and run with ~2 kW of electron beam power. The more expensive variety has a rotating-anode type source that run with ~14 kW of e-beam power.

X-rays are generally filtered (by use of X-Ray Filters) to a single wavelength (made monochromatic) and collimated to a single direction before they are allowed to strike the crystal. The filtering not only simplifies the data analysis, but also removes radiation that degrades the crystal without contributing useful information. Collimation is done either with a collimator (basically, a long tube) or with a clever arrangement of gently curved mirrors. Mirror systems are preferred for small crystals (under 0.3 mm) or with large unit cells (over 150 Å)

Synchrotron Radiation

Synchrotron radiation are some of the brightest lights on earth. It is the single most powerful tool available to X-ray crystallographers. It is made of X-ray beams generated in large machines called synchrotrons. These machines accelerate electrically charged particles, often electrons, to nearly the speed of light and confine them in a (roughly) circular loop using magnetic fields.

Synchrotrons are generally national facilities, each with several dedicated beamlines where data is collected without interruption. Synchrotrons were originally designed for use by high-energy physicists studying subatomic particles and cosmic phenomena. The largest component of each synchrotron is its electron storage ring. This ring is actually not a perfect circle, but a many-sided polygon. At each corner of the polygon, or sector, precisely aligned magnets bend the electron stream. As the electrons' path is bent, they emit bursts of energy in the form of X-rays.

Using synchrotron radiation frequently has specific requirements for X-ray crystallography. The intense ionizing radiation can cause radiation damage to samples, particularly macromolecular crystals. Cryo crystallography protects the sample from radiation damage, by freezing the crystal at liquid nitrogen temperatures (~100 K). However, synchrotron radiation frequently has the advantage of user selectable wavelengths, allowing for anomalous scattering experiments which maximizes anomalous signal. This is critical in experiments such as SAD and MAD.

Free Electron Laser

Recently, free electron lasers have been developed for use in X-ray crystallography. These are the brightest X-ray sources currently available; with the X-rays coming in femtosecond bursts. The intensity of the source is such that atomic resolution diffraction patterns can be resolved for crystals otherwise too small for collection. However, the intense light source also destroys the sample, requiring multiple crystals to be shot. As each crystal is randomly oriented in the beam, hundreds of thousands of individual diffraction images must be collected in order to get a complete data-set. This method, serial femtosecond crystallography, has been used in solving the structure of a number of protein crystal structures, sometimes noting differences with equivalent structures collected from synchrotron sources.

Recording the Reflections

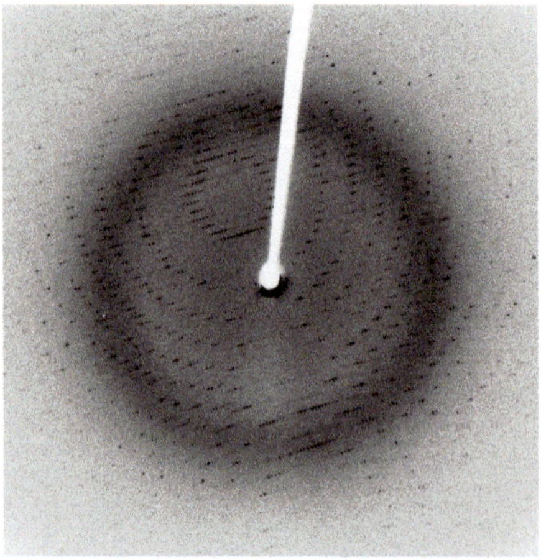

An X-ray diffraction pattern of a crystallized enzyme. The pattern of spots (*reflections*) and the relative strength of each spot (*intensities*) can be used to determine the structure of the enzyme.

When a crystal is mounted and exposed to an intense beam of X-rays, it scatters the X-rays into a pattern of spots or *reflections* that can be observed on a screen behind the crystal. A similar pattern may be seen by shining a laser pointer at a compact disc. The relative intensities of these spots provide the information to determine the arrangement of molecules within the crystal in atomic detail. The intensities of these reflections may be recorded with photographic film, an area detector or with a charge-coupled device (CCD) image sensor. The peaks at small angles correspond to low-resolution data, whereas those at high angles represent high-resolution data; thus, an upper limit on the eventual resolution of the structure can be determined from the first few images. Some measures of diffraction quality can be determined at this point, such as the mosaicity of the crystal and its overall disorder, as observed in the peak widths. Some pathologies of the crystal that would render it unfit for solving the structure can also be diagnosed quickly at this point.

One image of spots is insufficient to reconstruct the whole crystal; it represents only a small slice of the full Fourier transform. To collect all the necessary information, the crystal must be rotated step-by-step through 180°, with an image recorded at every step; actually, slightly more than 180° is required to cover reciprocal space, due to the curvature of the Ewald sphere. However, if the crystal has a higher symmetry, a smaller angular range such as 90° or 45° may be recorded. The rotation axis should be changed at least once, to avoid developing a "blind spot" in reciprocal space close to the rotation axis. It is customary to rock the crystal slightly (by 0.5–2°) to catch a broader region of reciprocal space.

Multiple data sets may be necessary for certain phasing methods. For example, MAD phasing requires that the scattering be recorded at least three (and usually four, for redundancy) wavelengths of the incoming X-ray radiation. A single crystal may degrade too much during the collection of one data set, owing to radiation damage; in such cases, data sets on multiple crystals must be taken.

Data Analysis

Crystal Symmetry, Unit Cell, and Image Scaling

The recorded series of two-dimensional diffraction patterns, each corresponding to a different crystal orientation, is converted into a three-dimensional model of the electron density; the conversion uses the mathematical technique of Fourier transforms, which is explained below. Each spot corresponds to a different type of variation in the electron density; the crystallographer must determine *which* variation corresponds to *which* spot (*indexing*), the relative strengths of the spots in different images (*merging and scaling*) and how the variations should be combined to yield the total electron density (*phasing*).

Data processing begins with *indexing* the reflections. This means identifying the dimensions of the unit cell and which image peak corresponds to which position in reciprocal space. A byproduct of indexing is to determine the symmetry of the crystal, i.e., its *space group*. Some space groups can be eliminated from the beginning. For example, reflection symmetries cannot be observed in chiral molecules; thus, only 65 space groups of 230 possible are allowed for protein molecules which are almost always chiral. Indexing is generally accomplished using an *autoindexing* routine. Having assigned symmetry, the data is then *integrated*. This converts the hundreds of images containing the thousands of reflections into a single file, consisting of (at the very least) records of the Miller index of each reflection, and an intensity for each reflection (at this state the file often also includes error estimates and measures of partiality (what part of a given reflection was recorded on that image)).

A full data set may consist of hundreds of separate images taken at different orientations of the crystal. The first step is to merge and scale these various images, that is, to identify which peaks appear in two or more images (*merging*) and to scale the relative images so that they have a consistent intensity scale. Optimizing the intensity scale is critical because the relative intensity of the peaks is the key information from which the structure is determined. The repetitive technique of crystallographic data collection and the often high symmetry of crystalline materials cause the diffractometer to record many symmetry-equivalent reflections multiple times. This allows calculating the symmetry-related R-factor, a reliability index based upon how similar are the measured intensities of symmetry-equivalent reflections, thus assessing the quality of the data.

Initial Phasing

The data collected from a diffraction experiment is a reciprocal space representation of the crystal lattice. The position of each diffraction 'spot' is governed by the size and shape of the unit cell, and the inherent symmetry within the crystal. The intensity of each diffraction 'spot' is recorded, and this intensity is proportional to the square of the *structure factor* amplitude. The structure factor is a complex number containing information relating to both the amplitude and phase of a wave. In order to obtain an interpretable *electron density map*, both amplitude and phase must be known (an electron density map allows a crystallographer to build a starting model of the molecule). The phase cannot be directly recorded during a diffraction experiment: this is known as the phase problem. Initial phase estimates can be obtained in a variety of ways:

- *Ab initio* phasing or direct methods – This is usually the method of choice for small molecules (<1000 non-hydrogen atoms), and has been used successfully to solve the phase problems for small proteins. If the resolution of the data is better than 1.4 Å (140 pm), direct methods can be used to obtain phase information, by exploiting known phase relationships between certain groups of reflections.

- Molecular replacement – if a related structure is known, it can be used as a search model in molecular replacement to determine the orientation and position of the molecules within the unit cell. The phases obtained this way can be used to generate *electron density maps*.
- Anomalous X-ray scattering (*MAD or SAD phasing*) – the X-ray wavelength may be scanned past an absorption edge of an atom, which changes the scattering in a known way. By recording full sets of reflections at three different wavelengths (far below, far above and in the middle of the absorption edge) one can solve for the substructure of the anomalously diffracting atoms and hence the structure of the whole molecule. The most popular method of incorporating anomalous scattering atoms into proteins is to express the protein in a methionine auxotroph (a host incapable of synthesizing methionine) in a media rich in seleno-methionine, which contains selenium atoms. A MAD experiment can then be conducted around the absorption edge, which should then yield the position of any methionine residues within the protein, providing initial phases.
- Heavy atom methods (multiple isomorphous replacement) – If electron-dense metal atoms can be introduced into the crystal, direct methods or Patterson-space methods can be used to determine their location and to obtain initial phases. Such heavy atoms can be introduced either by soaking the crystal in a heavy atom-containing solution, or by co-crystallization (growing the crystals in the presence of a heavy atom). As in MAD phasing, the changes in the scattering amplitudes can be interpreted to yield the phases. Although this is the original method by which protein crystal structures were solved, it has largely been superseded by MAD phasing with selenomethionine.

Model Building and Phase Refinement

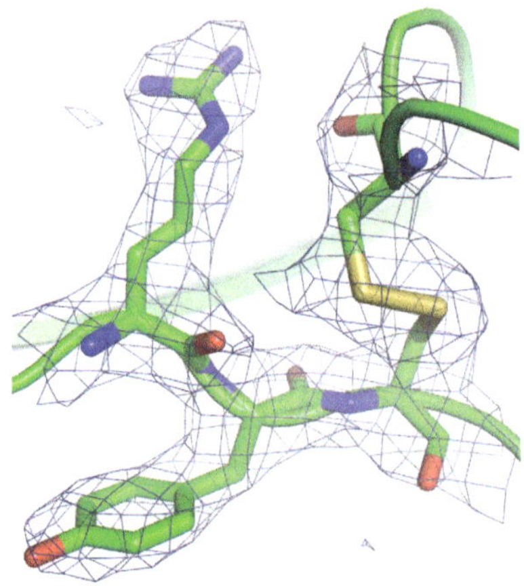

A protein crystal structure at 2.7 Å resolution. The mesh encloses the region in which the electron density exceeds a given threshold. The straight segments represent chemical bonds between the non-hydrogen atoms of an arginine (upper left), a tyrosine (lower left), a disulfide bond (upper right, in yellow), and some peptide groups (running left-right in the middle). The two curved green tubes represent spline fits to the polypeptide backbone.

Having obtained initial phases, an initial model can be built. This model can be used to refine the phases, leading to an improved model, and so on. Given a model of some atomic positions, these positions and their respective Debye-Waller factors (or B-factors, accounting for the thermal motion of the atom) can be refined to fit the observed diffraction data, ideally yielding a better set of phases. A new model can then be fit to the new electron density map and a further round of refinement is carried out. This continues until the correlation between the diffraction data and the model is maximized. The agreement is measured by an R-factor defined as

$$R = \frac{\sum_{all\ reflections} |F_o - F_c|}{\sum_{all\ reflections} |F_o|}$$

where F is the structure factor. A similar quality criterion is R_{free}, which is calculated from a subset (~10%) of reflections that were not included in the structure refinement. Both R factors depend on the resolution of the data. As a rule of thumb, R_{free} should be approximately the resolution in angstroms divided by 10; thus, a data-set with 2 Å resolution should yield a final R_{free} ~ 0.2. Chemical bonding features such as stereochemistry, hydrogen bonding and distribution of bond lengths and angles are complementary measures of the model quality. Phase bias is a serious problem in such iterative model building. *Omit maps* are a common technique used to check for this.

It may not be possible to observe every atom in the asymmetric unit. In many cases, disorder smears the electron density map. Weakly scattering atoms such as hydrogen are routinely invisible. It is also possible for a single atom to appear multiple times in an electron density map, e.g., if a protein sidechain has multiple (<4) allowed conformations. In still other cases, the crystallographer may detect that the covalent structure deduced for the molecule was incorrect, or changed. For example, proteins may be cleaved or undergo post-translational modifications that were not detected prior to the crystallization.

Disorder

A common challenge in refinement of crystal structures results from crystallographic disorder. Disorder can take many forms but in general involves the coexistence of two or more species or conformations. Failure to recognize disorder results in flawed interpretation. Pitfalls from improper modeling of disorder are illustrated by the discounted hypothesis of bond stretch isomerism. Disorder is modelled with respect to the relative population of the components, often only two, and their identity. In structures of large molecules and ions, solvent and counterions are often disordered.

Deposition of the Structure

Once the model of a molecule's structure has been finalized, it is often deposited in a crystallographic database such as the Cambridge Structural Database (for small molecules), the Inorganic Crystal Structure Database (ICSD) (for inorganic compounds) or the Protein Data Bank (for protein structures). Many structures obtained in private commercial ventures to crystallize medicinally relevant proteins are not deposited in public crystallographic databases.

Diffraction Theory

The main goal of X-ray crystallography is to determine the density of electrons $f(\mathbf{r})$ throughout the crystal, where $\mathbf{r}$ represents the three-dimensional position vector within the crystal. To do this, X-ray scattering is used to collect data about its Fourier transform $F(q)$, which is inverted mathematically to obtain the density defined in real space, using the formula

$$f(\mathbf{r}) = \frac{1}{(2\pi)^3} \int F(\mathbf{q}) e^{i\mathbf{q}\cdot\mathbf{r}} d\mathbf{q}$$

where the integral is taken over all values of q. The three-dimensional real vector q represents a point in reciprocal space, that is, to a particular oscillation in the electron density as one moves in the direction in which q points. The length of q corresponds to 2π divided by the wavelength of the oscillation. The corresponding formula for a Fourier transform will be used below

$$F(\mathbf{q}) = \int f(\mathbf{r}) e^{-i\mathbf{q}\cdot\mathbf{r}} d\mathbf{r}$$

where the integral is summed over all possible values of the position vector **r** within the crystal.

The Fourier transform F(q) is generally a complex number, and therefore has a magnitude $|F(\mathrm{q})|$ and a phase φ(q) related by the equation

$$F(\mathbf{q}) = |F(\mathbf{q})| e^{i\phi(\mathbf{q})}$$

The intensities of the reflections observed in X-ray diffraction give us the magnitudes $|F(\mathbf{q})|$ but not the phases φ(q). To obtain the phases, full sets of reflections are collected with known alterations to the scattering, either by modulating the wavelength past a certain absorption edge or by adding strongly scattering (i.e., electron-dense) metal atoms such as mercury. Combining the magnitudes and phases yields the full Fourier transform F(q), which may be inverted to obtain the electron density f(r).

Crystals are often idealized as being *perfectly* periodic. In that ideal case, the atoms are positioned on a perfect lattice, the electron density is perfectly periodic, and the Fourier transform $F(\mathbf{q})$ is zero except when q belongs to the reciprocal lattice (the so-called *Bragg peaks*). In reality, however, crystals are not perfectly periodic; atoms vibrate about their mean position, and there may be disorder of various types, such as mosaicity, dislocations, various point defects, and heterogeneity in the conformation of crystallized molecules. Therefore, the Bragg peaks have a finite width and there may be significant *diffuse scattering*, a continuum of scattered X-rays that fall between the Bragg peaks.

Intuitive Understanding by Bragg's Law

An intuitive understanding of X-ray diffraction can be obtained from the Bragg model of diffraction. In this model, a given reflection is associated with a set of evenly spaced sheets running through the crystal, usually passing through the centers of the atoms of the crystal lattice. The orientation of a particular set of sheets is identified by its three Miller indices (h, k, l), and let their spacing be noted by d. William Lawrence Bragg proposed a model in which the incoming X-rays are scattered specularly (mirror-like) from each plane; from that assumption, X-rays scattered from adjacent planes will combine constructively (constructive interference) when the angle θ between the plane and the X-ray results in a path-length difference that is an integer multiple n of the X-ray wavelength λ.

$$2d \sin\theta = n\lambda$$

A reflection is said to be *indexed* when its Miller indices (or, more correctly, its reciprocal lattice vector components) have been identified from the known wavelength and the scattering angle 2θ. Such indexing gives the unit-cell parameters, the lengths and angles of the unit-cell, as well as its space group. Since Bragg's law does not interpret the relative intensities of the reflections, however, it is generally inadequate to solve for the arrangement of atoms within the unit-cell; for that, a Fourier transform method must be carried out.

Scattering as a Fourier Transform

The incoming X-ray beam has a polarization and should be represented as a vector wave; however, for simplicity, let it be represented here as a scalar wave. We also ignore the complication of the time dependence of the wave and just concentrate on the wave's spatial dependence. Plane waves can be represented by a wave vector $\mathbf{k}_{in}$, and so the strength of the incoming wave at time $t=0$ is given by

$$Ae^{i\mathbf{k}_{in}\cdot\mathbf{r}}$$

At position **r** within the sample, let there be a density of scatterers $f(r)$; these scatterers should produce a scattered spherical wave of amplitude proportional to the local amplitude of the incoming wave times the number of scatterers in a small volume dV about r

$$amplitude\ of\ scattered\ wave = Ae^{i\mathbf{k}\cdot\mathbf{r}}Sf(\mathbf{r})dV$$

where S is the proportionality constant.

Let's consider the fraction of scattered waves that leave with an outgoing wave-vector of k_{out} and strike the screen at r_{screen}. Since no energy is lost (elastic, not inelastic scattering), the wavelengths are the same as are the magnitudes of the wave-vectors $|k_{in}|=|k_{out}|$. From the time that the photon is scattered at r until it is absorbed at r_{screen}, the photon undergoes a change in phase

$$e^{i\mathbf{k}_{out}\cdot(\mathbf{r}_{screen}-\mathbf{r})}$$

The net radiation arriving at r_{screen} is the sum of all the scattered waves throughout the crystal

$$AS\int d\mathbf{r}f(\mathbf{r})e^{i\mathbf{k}_{in}\cdot\mathbf{r}}e^{i\mathbf{k}_{out}\cdot(\mathbf{r}_{screen}-\mathbf{r})} = ASe^{i\mathbf{k}_{out}\cdot\mathbf{r}_{screen}}\int d\mathbf{r}f(\mathbf{r})e^{i(\mathbf{k}_{in}-\mathbf{k}_{out})\cdot\mathbf{r}}$$

which may be written as a Fourier transform

$$ASe^{i\mathbf{k}_{out}\cdot\mathbf{r}_{screen}}\int d\mathbf{r}f(\mathbf{r})e^{-i\mathbf{q}\cdot\mathbf{r}} = ASe^{i\mathbf{k}_{out}\cdot\mathbf{r}_{screen}}F(\mathbf{q})$$

where $q = k_{out} - k_{in}$. The measured intensity of the reflection will be square of this amplitude

$$A^2S^2|F(\mathbf{q})|^2$$

Friedel and Bijvoet Mates

For every reflection corresponding to a point q in the reciprocal space, there is another reflection of the same intensity at the opposite point -q. This opposite reflection is known as the *Friedel mate* of the original reflection. This symmetry results from the mathematical fact that the density of electrons $f(\mathbf{r})$ at a position r is always a real number. As noted above, $f(\mathrm{r})$ is the inverse transform of its Fourier transform $F(\mathrm{q})$; however, such an inverse transform is a complex number in general. To ensure that $f(\mathrm{r})$ is real, the Fourier transform $F(\mathrm{q})$ must be such that the Friedel mates $F(-\mathrm{q})$ and $F(\mathrm{q})$ are complex conjugates of one another. Thus, $F(-\mathrm{q})$ has the same magnitude as $F(\mathrm{q})$ but they have the opposite phase, i.e., $\varphi(\mathrm{q}) = -\varphi(\mathrm{q})$

$$F(-\mathbf{q}) = |F(-\mathbf{q})| e^{i\phi(-\mathbf{q})} = F^*(\mathbf{q}) = |F(\mathbf{q})| e^{-i\phi(\mathbf{q})}$$

The equality of their magnitudes ensures that the Friedel mates have the same intensity $|F|^2$. This symmetry allows one to measure the full Fourier transform from only half the reciprocal space, e.g., by rotating the crystal slightly more than 180° instead of a full 360° revolution. In crystals with significant symmetry, even more reflections may have the same intensity (Bijvoet mates); in such cases, even less of the reciprocal space may need to be measured. In favorable cases of high symmetry, sometimes only 90° or even only 45° of data are required to completely explore the reciprocal space.

The Friedel-mate constraint can be derived from the definition of the inverse Fourier transform

$$f(\mathbf{r}) = \int \frac{d\mathbf{q}}{(2\pi)^3} F(\mathbf{q}) e^{i\mathbf{q}\cdot\mathbf{r}} = \int \frac{d\mathbf{q}}{(2\pi)^3} |F(\mathbf{q})| e^{i\phi(\mathbf{q})} e^{i\mathbf{q}\cdot\mathbf{r}}$$

Since Euler's formula states that $e^{ix} = \cos(x) + i\sin(x)$, the inverse Fourier transform can be separated into a sum of a purely real part and a purely imaginary part

$$f(\mathbf{r}) = \int \frac{d\mathbf{q}}{(2\pi)^3} |F(\mathbf{q})| e^{i(\phi+\mathbf{q}\cdot\mathbf{r})} = \int \frac{d\mathbf{q}}{(2\pi)^3} |F(\mathbf{q})| \cos(\phi + \mathbf{q}\cdot\mathbf{r}) + i \int \frac{d\mathbf{q}}{(2\pi)^3} |F(\mathbf{q})| \sin(\phi + \mathbf{q}\cdot\mathbf{r}) = I_{\cos} + iI_{\sin}$$

The function $f(\mathbf{r})$ is real if and only if the second integral $I_{\sin}$ is zero for all values of $\mathbf{r}$. In turn, this is true if and only if the above constraint is satisfied

$$I_{\sin} = \int \frac{d\mathbf{q}}{(2\pi)^3} |F(\mathbf{q})| \sin(\phi + \mathbf{q}\cdot\mathbf{r}) = \int \frac{d\mathbf{q}}{(2\pi)^3} |F(-\mathbf{q})| \sin(-\phi - \mathbf{q}\cdot\mathbf{r}) = -I_{\sin}$$

since $I_{\sin} = -I_{\sin}$ implies that $I_{\sin} = 0$.

Ewald's Sphere

Each X-ray diffraction image represents only a slice, a spherical slice of reciprocal space, as may be seen by the Ewald sphere construction. Both k_{out} and k_{in} have the same length, due to the elastic scattering, since the wavelength has not changed. Therefore, they may be represented as two radial vectors in a sphere in reciprocal space, which shows the values of q that are sampled in a given diffraction image. Since there is a slight spread in the incoming wavelengths of the incoming X-ray beam, the values of $|F(\mathrm{q})|$ can be measured only for q vectors located between the two spheres corresponding to those radii. Therefore, to obtain a full set of Fourier transform data, it is necessary to rotate the crystal through slightly more than 180°, or sometimes less if sufficient symmetry is present. A full 360° rotation is not needed because of a symmetry intrinsic to the Fourier transforms of real functions (such as the electron density), but "slightly more" than 180° is needed to cover all of reciprocal space within a given resolution because of the curvature of the Ewald sphere. In practice, the crystal is rocked by a small amount (0.25-1°) to incorporate reflections near the boundaries of the spherical Ewald's shells.

Patterson Function

A well-known result of Fourier transforms is the autocorrelation theorem, which states that the autocorrelation *c*(r) of a function *f*(r)

$$c(\mathbf{r}) = \int d\mathbf{x} f(\mathbf{x}) f(\mathbf{x}+\mathbf{r}) = \int \frac{d\mathbf{q}}{(2\pi)^3} C(\mathbf{q}) e^{i\mathbf{q}\cdot\mathbf{r}}$$

has a Fourier transform *C*(q) that is the squared magnitude of *F*(q)

$$C(\mathbf{q}) = |F(\mathbf{q})|^2$$

Therefore, the autocorrelation function *c*(**r**) of the electron density (also known as the *Patterson function*) can be computed directly from the reflection intensities, without computing the phases. In principle, this could be used to determine the crystal structure directly; however, it is difficult to realize in practice. The autocorrelation function corresponds to the distribution of vectors between atoms in the crystal; thus, a crystal of *N* atoms in its unit cell may have *N(N-1)* peaks in its Patterson function. Given the inevitable errors in measuring the intensities, and the mathematical difficulties of reconstructing atomic positions from the interatomic vectors, this technique is rarely used to solve structures, except for the simplest crystals.

Advantages of a Crystal

In principle, an atomic structure could be determined from applying X-ray scattering to non-crystalline samples, even to a single molecule. However, crystals offer a much stronger signal due to their periodicity. A crystalline sample is by definition periodic; a crystal is composed of many unit cells repeated indefinitely in three independent directions. Such periodic systems have a Fourier transform that is concentrated at periodically repeating points in reciprocal space known as *Bragg peaks*; the Bragg peaks correspond to the reflection spots observed in the diffraction image. Since the amplitude at these reflections grows linearly with the number *N* of scatterers, the observed *intensity* of these spots should grow quadratically, like N^2. In other words, using a crystal concentrates the weak scattering of the individual unit cells into a much more powerful, coherent reflection that can be observed above the noise. This is an example of constructive interference.

In a liquid, powder or amorphous sample, molecules within that sample are in random orientations. Such samples have a continuous Fourier spectrum that uniformly spreads its amplitude thereby reducing the measured signal intensity, as is observed in SAXS. More importantly, the orientational information is lost. Although theoretically possible, it is experimentally difficult to obtain atomic-resolution structures of complicated, asymmetric molecules from such rotationally averaged data. An intermediate case is fiber diffraction in which the subunits are arranged periodically in at least one dimension.

Nobel Prizes for X-ray Crystallography

Year	Laureate	Prize	Rationale
1914	Max von Laue	Physics	"For his discovery of the diffraction of X-rays by crystals", an important step in the development of X-ray spectroscopy.
1915	William Henry Bragg	Physics	"For their services in the analysis of crystal structure by means of X-rays",
1915	William Lawrence Bragg	Physics	"For their services in the analysis of crystal structure by means of X-rays",
1962	Max F. Perutz	Chemistry	"for their studies of the structures of globular proteins"
1962	John C. Kendrew	Chemistry	"for their studies of the structures of globular proteins"
1962	James Dewey Watson	Medicine	"For their discoveries concerning the molecular structure of nucleic acids and its significance for information transfer in living material"
1962	Francis Harry Compton Crick	Medicine	"For their discoveries concerning the molecular structure of nucleic acids and its significance for information transfer in living material"
1962	Maurice Hugh Frederick Wilkins	Medicine	"For their discoveries concerning the molecular structure of nucleic acids and its significance for information transfer in living material"
1964	Dorothy Hodgkin	Chemistry	"For her determinations by X-ray techniques of the structures of important biochemical substances"
1972	Stanford Moore	Chemistry	"For their contribution to the understanding of the connection between chemical structure and catalytic activity of the active centre of the ribonuclease molecule"
1972	William H. Stein	Chemistry	"For their contribution to the understanding of the connection between chemical structure and catalytic activity of the active centre of the ribonuclease molecule"
1976	William N. Lipscomb	Chemistry	"For his studies on the structure of boranes illuminating problems of chemical bonding"
1985	Jerome Karle	Chemistry	"For their outstanding achievements in developing direct methods for the determination of crystal structures"
1985	Herbert A. Hauptman	Chemistry	"For their outstanding achievements in developing direct methods for the determination of crystal structures"
1988	Johann Deisenhofer	Chemistry	"For their determination of the three-dimensional structure of a photosynthetic reaction centre"
1988	Hartmut Michel	Chemistry	"For their determination of the three-dimensional structure of a photosynthetic reaction centre"
1988	Robert Huber	Chemistry	"For their determination of the three-dimensional structure of a photosynthetic reaction centre"
1997	John E. Walker	Chemistry	"For their elucidation of the enzymatic mechanism underlying the synthesis of adenosine triphosphate (ATP)"
2003	Roderick MacKinnon	Chemistry	"For discoveries concerning channels in cell membranes [...] for structural and mechanistic studies of ion channels"
2003	Peter Agre	Chemistry	"For discoveries concerning channels in cell membranes [...] for the discovery of water channels"
2006	Roger D. Kornberg	Chemistry	"For his studies of the molecular basis of eukaryotic transcription"
2009	Ada E. Yonath	Chemistry	"For studies of the structure and function of the ribosome"

2009	Thomas A. Steitz	Chemistry	"For studies of the structure and function of the ribosome"
2009	Venkatraman Ramakrishnan	Chemistry	"For studies of the structure and function of the ribosome"
2012	Brian Kobilka	Chemistry	"For studies of G-protein-coupled receptors"

Applications of X-ray Diffraction

X-ray diffraction has a wide and various applications on the chemical, biochemical, physical, material and mineralogical sciences. Laue`s said that 'has extended the power of on serving minute structure ten thousand times beyond that of the optical microscope. X-ray diffraction produced a microscope with atomic resolution which shows the atoms and their electron distribution. X-ray diffraction, electron diffraction, and neutron diffraction give information about the structure of matter, crystalline and non-crystalline, at the atomic and molecular level. In addition, they are attached to the properties of all materials, inorganic, organic or biological. Due to the diffraction importance and variety of application of diffraction by crystals, a large number of Nobel Prizes were presented to studies involving X-ray.

X-ray Method for Investigation of Drugs

X-ray diffraction was used for the identification of antibiotic drugs such as: eight β-lactam (ampicillin sodium, penicillin G procaine, cefalexin, ampicillin trihydrate, benzathine penicillin, benzylpenicillin sodium, cefotaxime sodium, Ceftriaxone sodium), three tetracycline (doxycycline hydrochloride, oxytetracycline dehydrate, tetracycline hydrochloride) and two macrolide (azithromycin, erythromycin estolate) antibiotic drugs. Each of these drugs has a unique XRD pattern that makes their identification possible.

X-ray Method for Investigation of Textile Fibers and Polymers

Forensic examination of any trace evidence is based upon Locard's exchange principle. This states that *every contact leaves a trace*. In practice, even though a transfer of material has taken place, it may be impossible to detect, because the amount transferred is very small. Textile fibers are a mixture of crystalline and amorphous substances. Therefore the measurement of the degree of crystalline gives useful data in the characterization of fibers using X-ray diffractometry. It has been reported that X-ray diffraction was used to identify of a "crystalline" deposit which was found on a chair. The deposit was found to be amorphous, but the diffraction pattern present matched that of polymethylmethacrylate. Pyrolysis mass spectrometry later identified the deposit as polymethylcyanoacrylaon of Boin crystal parameters.

X-ray Method for Investigation of Bones

Hiller investigated the effects of heating and burning on bone mineral using X-ray diffraction (XRD) techniques. The bone samples were heated in temperature of 500, 700, and 900 C° for 15 and 45 min. The results show bone crystals began to change during the first 15 min of heating at 500 Co and above. At higher temperatures, thickness and shape of crystals of bones appear stabilized, but when the samples were heated at lower temperature or for shorter period, XRD traces showed extreme changes in crystal parameters.

References

- Garman, E. F.; Schneider, T. R. (1997). "Macromolecular Cryocrystallography". Journal of Applied Crystallography. 30 (3): 211. doi:10.1107/S0021889897002677
- Schwartz, Mel (2002). "Tin and Alloys, Properties". Encyclopedia of Materials, Parts and Finishes (2nd ed.). CRC Press. ISBN 1-56676-661-3
- Snigirev, A. (2007). "Two-step hard X-ray focusing combining Fresnel zone plate and single-bounce ellipsoidal capillary". Journal of Synchrotron Radiation. 14 (Pt 4): 326–330. PMID 17587657. doi:10.1107/S0909049507025174
- "4. Direct and reciprocal lattices". CSIC Dept de Cristalografia y Biologia Estructural. 6 Apr 2017. Retrieved 18 May 2017.
- L. Pauling (1929). "The principles determining the structure of complex ionic crystals". J. Am. Chem. Soc. 51 (4): 1010–1026. doi:10.1021/ja01379a006
- Holleman, Arnold F.; Wiberg, Egon; Wiberg, Nils (1985). "Tin". Lehrbuch der Anorganischen Chemie (in German) (91–100 ed.). Walter de Gruyter. pp. 793–800. ISBN 3-11-007511-3
- Bernal, J. D.; Fowler, R. H. (1933). "A Theory of Water and Ionic Solution, with Particular Reference to Hydrogen and Hydroxyl Ions". The Journal of Chemical Physics. 1 (8): 515. Bibcode:1933JChPh...1..515B. doi:10.1063/1.1749327
- Von Laue M (1914). "Concerning the detection of x-ray interferences" (PDF). Nobel Lectures, Physics. 1901–1921. Retrieved 2009-02-18
- Bragg WL (1912). "The Specular Reflexion of X-rays". Nature. 90 (2250): 410. Bibcode:1912Natur..90..410B. doi:10.1038/090410b0
- Ashcroft, N.; Mermin, D. (1976). "Chapter 7". Solid State Physics. Brooks/Cole (Thomson Learning, Inc.). ISBN 0030493463
- Pauling, Linus (1947). "Atomic Radii and Interatomic Distances in Metals". Journal of the American Chemical Society. 69 (3): 542. doi:10.1021/ja01195a024
- "From Atoms To Patterns". Wellcome Collection. Archived from the original on September 7, 2013. Retrieved 17 October 2013
- Wunderlich JA; Mellor DP (1954). "A note on the crystal structure of Zeise's salt". Acta Crystallographica. 7: 130. doi:10.1107/S0365110X5400028X
- Hook, J.R.; Hall, H.E. (2010). Solid State Physics. Manchester Physics Series (2nd ed.). John Wiley & Sons. ISBN 9780471928041
- Saville WB; Shearer G (1925). "An X-ray Investigation of Saturated Aliphatic Ketones". Journal of the Chemical Society (London). 127: 591. doi:10.1039/ct9252700591
- Brown, Dwayne (October 30, 2012). "NASA Rover's First Soil Studies Help Fingerprint Martian Minerals". NASA. Retrieved October 31, 2012
- Ealick SE (2000). "Advances in multiple wavelength anomalous diffraction crystallography". Current Opinion in Chemical Biology. 4 (5): 495–9. PMID 11006535. doi:10.1016/S1367-5931(00)00122-8
- Jeruzalmi D (2006). "First analysis of macromolecular crystals: biochemistry and x-ray diffraction". Methods Mol. Biol. 364: 43–62. ISBN 1-59745-266-1. PMID 17172760. doi:10.1385/1-59745-266-1:43

Permissions

All chapters in this book are published with permission under the Creative Commons Attribution Share Alike License or equivalent. Every chapter published in this book has been scrutinized by our experts. Their significance has been extensively debated. The topics covered herein carry significant information for a comprehensive understanding. They may even be implemented as practical applications or may be referred to as a beginning point for further studies.

We would like to thank the editorial team for lending their expertise to make the book truly unique. They have played a crucial role in the development of this book. Without their invaluable contributions this book wouldn't have been possible. They have made vital efforts to compile up to date information on the varied aspects of this subject to make this book a valuable addition to the collection of many professionals and students.

This book was conceptualized with the vision of imparting up-to-date and integrated information in this field. To ensure the same, a matchless editorial board was set up. Every individual on the board went through rigorous rounds of assessment to prove their worth. After which they invested a large part of their time researching and compiling the most relevant data for our readers.

The editorial board has been involved in producing this book since its inception. They have spent rigorous hours researching and exploring the diverse topics which have resulted in the successful publishing of this book. They have passed on their knowledge of decades through this book. To expedite this challenging task, the publisher supported the team at every step. A small team of assistant editors was also appointed to further simplify the editing procedure and attain best results for the readers.

Apart from the editorial board, the designing team has also invested a significant amount of their time in understanding the subject and creating the most relevant covers. They scrutinized every image to scout for the most suitable representation of the subject and create an appropriate cover for the book.

The publishing team has been an ardent support to the editorial, designing and production team. Their endless efforts to recruit the best for this project, has resulted in the accomplishment of this book. They are a veteran in the field of academics and their pool of knowledge is as vast as their experience in printing. Their expertise and guidance has proved useful at every step. Their uncompromising quality standards have made this book an exceptional effort. Their encouragement from time to time has been an inspiration for everyone.

The publisher and the editorial board hope that this book will prove to be a valuable piece of knowledge for students, practitioners and scholars across the globe.

Index